FREE

Free Study Tips DVD

In addition to the tips and content in this guide, we have created a FREE DVD with helpful study tips to further assist your exam preparation. **This FREE Study Tips DVD provides you with top-notch tips to conquer your exam and reach your goals.**

Our simple request in exchange for the strategy-packed DVD is that you email us your feedback about our study guide. We would love to hear what you thought about the guide, and we welcome any and all feedback—positive, negative, or neutral. It is our #1 goal to provide you with top quality products and customer service.

To receive your **FREE Study Tips DVD**, email freedvd@apexprep.com. Please put "FREE DVD" in the subject line and put the following in the email:

a. The name of the study guide you purchased.

b. Your rating of the study guide on a scale of 1-5, with 5 being the highest score.

c. Any thoughts or feedback about your study guide.

d. Your first and last name and your mailing address, so we know where to send your free DVD!

Thank you!

GRE Prep 2022 and 2023

GRE Study Book with Practice Test Questions [7th Edition]

Matthew Lanni

Written and edited by APEX Publishing.

ISBN 13: 9781637757567
ISBN 10: 1637757565

For additional information or for bulk orders, contact info@apexprep.com.

Table of Contents

Test Taking Strategies

1. Reading the Whole Question

A popular assumption in Western culture is the idea that we don't have enough time for anything. We speed while driving to work, we want to read an assignment for class as quickly as possible, or we want the line in the supermarket to dwindle faster. However, speeding through such events robs us from being able to thoroughly appreciate and understand what's happening around us. While taking a timed test, the feeling one might have while reading a question is to find the correct answer as quickly as possible. Although pace is important, don't let it deter you from reading the whole question. Test writers know how to subtly change a test question toward the end in various ways, such as adding a negative or changing focus. If the question has a passage, carefully read the whole passage as well before moving on to the questions. This will help you process the information in the passage rather than worrying about the questions you've just read and where to find them. A thorough understanding of the passage or question is an important way for test takers to be able to succeed on an exam.

2. Examining Every Answer Choice

Let's say we're at the market buying apples. The first apple we see on top of the heap may *look* like the best apple, but if we turn it over we can see bruising on the skin. We must examine several apples before deciding which apple is the best. Finding the correct answer choice is like finding the best apple. Although it's tempting to choose an answer that seems correct at first without reading the others, it's important to read each answer choice thoroughly before making a final decision on the answer. The aim of a test writer might be to get as close as possible to the correct answer, so watch out for subtle words that may indicate an answer is incorrect. Once the correct answer choice is selected, read the question again and the answer in response to make sure all your bases are covered.

3. Eliminating Wrong Answer Choices

Sometimes we become paralyzed when we are confronted with too many choices. Which frozen yogurt flavor is the tastiest? Which pair of shoes look the best with this outfit? What type of car will fill my needs as a consumer? If you are unsure of which answer would be the best to choose, it may help to use process of elimination. We use "filtering" all the time on sites such as eBay® or Craigslist® to eliminate the ads that are not right for us. We can do the same thing on an exam. Process of elimination is crossing out the answer choices we know for sure are wrong and leaving the ones that might be correct. It may help to cover up the incorrect answer choice. Covering incorrect choices is a psychological act that alleviates stress due to the brain being exposed to a smaller amount of information. Choosing between two answer choices is much easier than choosing between all of them, and you have a better chance of selecting the correct answer if you have less to focus on.

4. Sticking to the World of the Question

When we are attempting to answer questions, our minds will often wander away from the question and what it is asking. We begin to see answer choices that are true in the real world instead of true in the world of the question. It may be helpful to think of each test question as its own little world. This world may be different from ours. This world may know as a truth that the chicken came before the egg or may assert that two plus two equals five. Remember that, no matter what hypothetical nonsense may be in the question, assume it to be true. If the question states that the chicken came before the egg, then choose your answer based on that truth. Sticking to the world of the question means placing all of our biases and assumptions aside and relying on the question to guide us to the correct answer. If we are simply looking for answers that are correct based on our own judgment, then we may choose incorrectly. Remember an answer that is true does not necessarily answer the question.

5. Key Words

If you come across a complex test question that you have to read over and over again, try pulling out some key words from the question in order to understand what exactly it is asking. Key words may be words that surround the question, such as *main idea, analogous, parallel, resembles, structured,* or *defines.* The question may be asking for the main idea, or it may be asking you to define something. Deconstructing the sentence may also be helpful in making the question simpler before trying to answer it. This means taking the sentence apart and obtaining meaning in pieces, or separating the question from the foundation of the question. For example, let's look at this question:

> Given the author's description of the content of paleontology in the first paragraph, which of the following is most parallel to what it taught?

The question asks which one of the answers most *parallels* the following information: The *description* of paleontology in the first paragraph. The first step would be to see *how* paleontology is described in the first paragraph. Then, we would find an answer choice that parallels that description. The question seems complex at first, but after we deconstruct it, the answer becomes much more attainable.

6. Subtle Negatives

Negative words in question stems will be words such as *not, but, neither,* or *except.* Test writers often use these words in order to trick unsuspecting test takers into selecting the wrong answer—or, at least, to test their reading comprehension of the question. Many exams will feature the negative words in all caps (*which of the following is NOT an example*), but some questions will add the negative word seamlessly into the sentence. The following is an example of a subtle negative used in a question stem:

> According to the passage, which of the following is *not* considered to be an example of paleontology?

If we rush through the exam, we might skip that tiny word, *not,* inside the question, and choose an answer that is opposite of the correct choice. Again, it's important to read the question fully, and double check for any words that may negate the statement in any way.

7. Spotting the Hedges

The word "hedging" refers to language that remains vague or avoids absolute terminology. Absolute terminology consists of words like *always, never, all, every, just, only, none,* and *must.* Hedging refers to words like *seem, tend, might, most, some, sometimes, perhaps, possibly, probability,* and *often.* In some cases, we want to choose answer choices that use hedging and avoid answer choices that use absolute terminology. It's important to pay attention to what subject you are on and adjust your response accordingly.

8. Restating to Understand

Every now and then we come across questions that we don't understand. The language may be too complex, or the question is structured in a way that is meant to confuse the test taker. When you come across a question like this, it may be worth your time to rewrite or restate the question in your own words in order to understand it better. For example, let's look at the following complicated question:

> Which of the following words, if substituted for the word *parochial* in the first paragraph, would LEAST change the meaning of the sentence?

Let's restate the question in order to understand it better. We know that they want the word *parochial* replaced. We also know that this new word would "least" or "not" change the meaning of the sentence. Now let's try the sentence again:

> Which word could we replace with *parochial*, and it would not change the meaning?

Restating it this way, we see that the question is asking for a synonym. Now, let's restate the question so we can answer it better:

> Which word is a synonym for the word *parochial*?

Before we even look at the answer choices, we have a simpler, restated version of a complicated question.

9. Predicting the Answer

After you read the question, try predicting the answer *before* reading the answer choices. By formulating an answer in your mind, you will be less likely to be distracted by any wrong answer choices. Using predictions will also help you feel more confident in the answer choice you select. Once you've chosen your answer, go back and reread the question and answer choices to make sure you have the best fit. If you have no idea what the answer may be for a particular question, forego using this strategy.

10. Avoiding Patterns

One popular myth in grade school relating to standardized testing is that test writers will often put multiple-choice answers in patterns. A runoff example of this kind of thinking is that the most common answer choice is "C," with "B" following close behind. Or, some will advocate certain made-up word patterns that simply do not exist. Test writers do not arrange their correct answer choices in any kind of pattern; their choices are randomized. There may even be times where the correct answer choice will be the same letter for two or three questions in a row, but we have no way of knowing when or if this might happen. Instead of trying to figure out what choice the test writer probably set as being correct, focus on what the *best answer choice* would be out of the answers you are presented with. Use the tips above, general knowledge, and reading comprehension skills in order to best answer the question, rather than looking for patterns that do not exist.

FREE DVD OFFER

Achieving a high score on your exam depends not only on understanding the content, but also on understanding how to apply your knowledge and your command of test taking strategies. **Because your success is our primary goal, we offer a FREE Study Tips DVD, which provides top-notch test taking strategies to help you optimize your testing experience.**

Our simple request in exchange for the strategy-packed DVD is that you email us your feedback about our study guide.

To receive your **FREE Study Tips DVD**, email freedvd@apexprep.com. Please put "FREE DVD" in the subject line and put the following in the email:

a. The name of the study guide you purchased.

b. Your rating of the study guide on a scale of 1-5, with 5 being the highest score.

c. Any thoughts or feedback about your study guide.

d. Your first and last name and your mailing address, so we know where to send your free DVD!

Introduction to the GRE

Function of the Test

Administered by the Educational Testing Service (ETS), the Graduate Record Examinations (GRE) is an exam required for admission in many university graduate-level programs. Students who take the GRE have finished their undergraduate degree and are looking to go to graduate school. The exam measures students' readiness for entrance into higher-level programs and uses the score to determine entrance into a specific program.

Scoring is important depending on what program the student is entering into. For example, a liberal arts program may only be interested in the student's verbal score of the GRE and ignore the quantitative score altogether. Sometimes the GRE score is important as a whole, and sometimes only a section is important, so check with your desired program to understand what score they are looking for. From July 2016 to June 2017, 559,254 people took the GRE outside and within the United States. The average Verbal Reasoning score was 150.4. The average Quantitative Reasoning score was 153.2. The average Analytical Writing score was 3.6.

Test Administration

The GRE is offered in more than 160 countries with over 1,000 testing centers. 99% of examinees take the computer-based test, but those who are in areas that only offer the paper-based test can take the GRE up to three times a year in October, November, and February. Within the United States, the GRE is available at Prometric testing centers year-round. Students can retake the exam once every 21 days up to five times a year. For disability accommodations, submit requests to ETS before scheduling a test date. Visit the examinee's electronic ETS account to learn about the process for requesting accommodations.

Test Format

Valid identification is required to enter the examination site, and examinees should arrive 30 minutes prior to the scheduled testing time. Food and beverages are not allowed in the testing room. Seats are assigned in the testing room, and test takers must ask permission to leave the test center building, otherwise scores may be forfeited without refund.

The Computer-based GRE takes 3 hours and 45 minutes to complete, with six sections and a 10-minute break following the third section. The exam measures Analytical Writing, Verbal Reasoning, and Quantitative Reasoning. Verbal Reasoning assesses reading comprehension, critical reasoning, and vocabulary usage. Analytical Writing assesses writing abilities. Quantitative Reasoning assesses quantitative comparisons, problem solving, and data interpretation.

Below is a table with more information on the various sections:

Section	Time	Description
Analytical Writing	60 minutes	2 tasks. One 30-minute "Analyze an Issue" task and one 30-minute "Analyze an Argument" task
Verbal Reasoning	60 minutes	40 questions (Two sections with 20 questions per section)
Quantitative Reasoning	70 minutes	40 questions (Two sections with 20 questions per section)
Unscored/Research	Varies	Varies

Note that the GRE is adaptive, allowing examinees to go back and change answers if they need to. Test takers can "mark" or "review" answers they want to revisit. An onscreen calculator is also available for the Quantitative Reasoning section.

Scoring

With the ETS ScoreSelect option, examinees are able to choose which test scores get sent to the program or institution of their choice. For tests taken after July 1, 2016, scores are reportable for five years after the testing date. Verbal Reasoning scores range from 130-170 in 1-point increments. Quantitative Reasoning scores range from 130-170 in 1-point increments. Analytical writing scores range from zero to 6 in half-point increments.

Recent/Future Developments

The most recent change to the GRE is the introduction of "GRE revised" in 2011, which changed the scoring from a 1600-point scale to the 130–170 point scale and made the exam adaptive from section to section rather than from question to question.

Study Prep Plan for the GRE

1 Breathe

Reducing stress is key when preparing for your test.

2 Build

Create a study plan to help you stay on track.

3 Begin

Stick with your study plan. You've got this!

1 Week Study Plan

Day 1	Day 2	Day 3	Day 4	Day 5	Day 6	Day 7
Verbal Reasoning	Practice Questions	Quantitative Reasoning	Practice Questions	Analytical Writing	Practice Questions	Take your exam!

2 Week Study Plan

Day 1	Day 2	Day 3	Day 4	Day 5	Day 6	Day 7
Drawing Conclusions from Discourse	Comparing and Contrasting Themes from Print and Other Sources	Practice Questions	Review Answer Explanations	Properties and Types of Integers	Arithmetic Concepts	Solving Simultaneous Equations
Day 8	**Day 9**	**Day 10**	**Day 11**	**Day 12**	**Day 13**	**Day 14**
Data Analysis	Practice Questions	Review Answer Explanations	Articulating Ideas Clearly and Effectively	Controlling the Elements of Standard Written English	Practice Questions	Take your exam!

30 Day Study Plan

Day 1	Day 2	Day 3	Day 4	Day 5	Day 6	Day 7
Drawing Conclusions from Discourse	Author's Intent	Evaluating the Author's Point of View in a Given Text	Recognizing Events in Sequence	Context Clues	Practice Questions	Review Answer Explanations

Day 8	Day 9	Day 10	Day 11	Day 12	Day 13	Day 14
Properties and Types of Integers	Arithmetic Operations	Exponents	Ratio/Rate	Operations with Exponents	Solving Equations	Solving Simultaneous Equations

Day 15	Day 16	Day 17	Day 18	Day 19	Day 20	Day 21
Coordinate Geometry	Geometry	Congruent and Similar Figures	Area	Pythagorean Theorem	Elementary Probability	Counting Methods

Day 22	Day 23	Day 24	Day 25	Day 26	Day 27	Day 28
Practice Questions	Review Answer Explanations	Articulating Ideas Clearly and Effectively	Examining Claims and Evidence	Referencing Sources	Developing a Well Organized Paragraph	Conventions of Standard English Punctuation

Day 29	Day 30
Practice Questions	Take your exam!

Verbal Reasoning

Drawing Conclusions from Discourse

An **inference** is an educated guess or conclusion based on sound evidence and reasoning within the text. The test may include multiple-choice questions asking about the logical conclusion that can be drawn from reading a text, and you will have to identify the choice that unavoidably leads to that conclusion. In order to eliminate the incorrect choices, the test-taker should come up with a hypothetical situation wherein an answer choice is true, but the conclusion is not true. Here is an example:

> Fred purchased the newest PC available on the market. Therefore, he purchased the most expensive PC in the computer store.
>
> What can one assume for this conclusion to follow logically?
>
> a. Fred enjoys purchasing expensive items.
> b. PCs are some of the most expensive personal technology products available.
> c. The newest PC is the most expensive one.

The premise of the text is the first sentence: Fred purchased the newest PC. The conclusion is the second sentence: Fred purchased the most expensive PC. Recent release and price are two different factors; the difference between them is the logical gap. To eliminate the gap, one must connect the new information from the conclusion with the pertinent information from the premise. In this example, there must be a connection between product recency and product price. Therefore, a possible bridge to the logical gap could be a sentence stating that the newest PCs always cost the most.

Reasoning from Incomplete Data

One technique authors often use to make their fictional stories more interesting is not giving away too much information by providing hints and description. It is then up to the reader to draw a conclusion about the author's meaning by connecting textual clues with the reader's own pre-existing experiences and knowledge. Drawing conclusions is important as a reading strategy for understanding what is occurring in a text. Rather than directly stating who, what, where, when, or why, authors often describe story elements. Then, readers must draw conclusions to understand significant story components. As they go through a text, readers can think about the setting, characters, plot, problem, and solution; whether the author provided any clues for consideration; and combine any story clues with their existing knowledge and experiences to draw conclusions about what occurs in the text.

Making Predictions

Before and during reading, readers can apply the reading strategy of making predictions about what they think may happen next. For example, what plot and character developments will occur in fiction? What points will the author discuss in nonfiction? Making predictions about portions of text they have not yet read prepares readers mentally for reading, and also gives them a purpose for reading.

To inform and make predictions about text, the reader can do the following:

- Consider the title of the text and what it implies.
- Look at the cover of the book.
- Look at any illustrations or diagrams for additional visual information.
- Analyze the structure of the text.
- Apply outside experience and knowledge to the text.

Readers may adjust their predictions as they read. Reader predictions may or may not come true in text.

Making Inferences

Authors describe settings, characters, character emotions, and events. Readers must infer to understand a text fully. Inferring enables readers to figure out meanings of unfamiliar words, make predictions about upcoming text, draw conclusions, and reflect on reading. Readers can infer about text before, during, and after reading. In everyday life, we use sensory information to infer. Readers can do the same with text. When authors do not answer all readers' questions, readers must infer by saying "I think....This could be....This is because....Maybe....This means....I guess..." etc. Looking at illustrations, considering characters' behaviors, and asking questions during reading facilitate inference. Taking clues from text and connecting text to prior knowledge helps to draw conclusions. Readers can infer word meanings, settings, reasons for occurrences, character emotions, pronoun referents, author messages, and answers to questions unstated in text. To practice inference, students can read sentences written/selected by the instructor, discuss the setting and characters, draw conclusions, and make predictions.

Making inferences and drawing conclusions involve skills that are quite similar: both require readers to fill in information the author has omitted. Authors may omit information as a technique for inducing readers to discover the outcomes themselves; or they may consider certain information unimportant; or they may assume their reading audience already knows certain information. To make an inference or draw a conclusion about text, readers should observe all facts and arguments the author has presented and consider what they already know from their own personal experiences. Reading students taking multiple-choice tests that refer to text passages can determine correct and incorrect choices based on the information in the passage. For example, from a text passage describing an individual's signs of anxiety while unloading groceries and nervously clutching their wallet at a grocery store checkout, readers can infer or conclude that the individual may not have enough money to pay for everything.

Author's Assumptions

Facts and Opinions

A **fact** is a statement that is true empirically or an event that has actually occurred in reality and can be proven or supported by evidence; it is generally objective. In contrast, an **opinion** is subjective, representing something that someone believes rather than something that exists in the absolute. People's individual understandings, feelings, and perspectives contribute to variations in opinion. Though facts are typically objective in nature, in some instances, a statement of fact may be both factual and yet also subjective. For example, emotions are individual subjective experiences. If an individual says that they feel happy or sad, the feeling is subjective, but the statement is factual; hence, it is a subjective fact. In contrast, if one person tells another that the other is feeling happy or sad—whether this is true or not—that is an assumption or an opinion.

Biases

Biases usually occur when someone allows their personal preferences or ideologies to interfere with what should be an objective decision. In personal situations, someone is biased towards someone if they favor them in an unfair way. In academic writing, being biased in your sources means leaving out objective information that would turn the argument one way or the other. The evidence of bias in academic writing makes the text less credible, so be sure to present all viewpoints when writing, not just your own, so to avoid coming off as biased. Being objective when presenting information or dealing with people usually allows the person to gain more credibility.

Stereotypes

Stereotypes are preconceived notions that place a particular rule or characteristics on an entire group of people. Stereotypes are usually offensive to the group they refer to or allies of that group and often have negative connotations. The reinforcement of stereotypes isn't always obvious. Sometimes stereotypes can be very subtle and are still widely used in order for people to understand categories within the world. For example, saying that women are more emotional and intuitive than men is a stereotype that is still used by many to explain differences between individuals, even though it is erroneous.

Literal and Figurative Language

Authors of a text use language with multiple levels of meaning for many different reasons. When the meaning of a text requires directness, the author will use **literal language** to provide clarity. On the other hand, an author will use **figurative language** to produce an emotional effect or facilitate a deeper understanding of a word or passage. For example, a set of instructions on how to use a computer would require literal language. However, a commentary on the social implications of immigration bans might contain a wide range of figurative language to elicit an empathetic response. A single text can have a mixture of both literal and figurative language.

Literal Language

Literal language uses words in accordance with their actual definition. Many informational texts employ literal language because it is straightforward and precise. Documents such as instructions, proposals, technical documents, and workplace documents use literal language for the majority of their writing, so there is no confusion or complexity of meaning for readers to decipher. The information is best communicated through clear and precise language. The following are brief examples of literal language:

- I cook with olive oil.
- There are 365 days in a year.
- My grandma's name is Barbara.
- Yesterday we had some scattered thunderstorms.
- World War II began in 1939.
- Blue whales are the largest species of whale.

Figurative Language

Not meant to be taken literally, **figurative language** is useful when the author of a text wants to produce an emotional effect in the reader or add a heightened complexity to the meaning of the text. Figurative language is used more heavily in texts such as literary fiction, poetry, critical theory, and speeches. It goes beyond literal language, allowing readers to form associations they wouldn't normally form. Using language in a figurative sense appeals to the imagination of the reader. It is important to remember that words signify objects and ideas and are not the objects and ideas themselves. Figurative language can highlight this detachment by creating multiple associations, but it also points to the fact that language is fluid and capable of creating a world full of linguistic possibilities. It can be argued that figurative language is the heart of communication even outside of fiction and poetry. People connect through humor, metaphors, cultural allusions, puns, and symbolism in their everyday rhetoric. The following are terms associated with figurative language:

Simile

A **simile** is a comparison of two things using *like*, *than*, or *as*. A simile usually takes objects that have no apparent connection, such as a mind and an orchid, and compares them:

> His mind was as complex and rare as a field of ghost orchids.

Similes encourage new, fresh perspectives on objects or ideas that would not otherwise occur. Unlike similes, **metaphors** are comparisons that do not use *like*, *than*, or *as*. So, a metaphor from the above example would be:

> His mind was a field of ghost orchids.

Thus, similes highlight the comparison by focusing on the figurative side of the language, elucidating the author's intent. Metaphors, however, provide a beautiful yet somewhat equivocal comparison.

Metaphor

A popular use of figurative language, **metaphors** compare objects or ideas directly, asserting that something *is* a certain thing, even if it isn't. The following is an example of a metaphor used by writer Virginia Woolf:

> Books are the mirrors of the soul.

Metaphors have a vehicle and a tenor. The tenor is "books" and the vehicle is "mirrors of the soul." That is, the tenor is what is meant to be described, and the vehicle is that which carries the weight of the comparison. In this metaphor, perhaps the author means to say that written language (books) reflect a person's most inner thoughts and desires.

Dead metaphors are phrases that have been overused to the point where the figurative language has taken on a literal meaning, like "crystal clear." This phrase is in such popular use that the meaning seems literal ("perfectly clear") even when it is not.

Finally, an **extended metaphor** is one that goes on for several paragraphs,, or even an entire text. "On First Looking into Chapman's Homer," a poem by John Keats, begins, "Much have I travell'd in the realms of gold," and goes on to explain the first time he hears Chapman's translation of Homer's writing. We see the extended metaphor begin in the first line. Keats is comparing travelling into "realms of gold" and exploration of new lands to the act of hearing a certain kind of literature for the first time. The extended metaphor goes on until the end of the poem where Keats stands "Silent, upon a peak in Darien," having heard the end of Chapman's translation. Keats has gained insight into new lands (new text) and is the richer for it.

The following are brief definitions and examples of popular figurative language:

Onomatopoeia: A word that, when spoken, imitates the sound to which it refers. Ex: "We heard a loud *boom* while driving to the beach yesterday."

Personification: When human characteristics are given to animals, inanimate objects, or abstractions. An example would be in William Wordsworth's poem "Daffodils" where he sees a "crowd . . . / of golden daffodils . . . / Fluttering and dancing in the breeze." Dancing is usually a characteristic attributed solely to humans, but Wordsworth personifies the daffodils here as a crowd of people dancing.

Juxtaposition: Juxtaposition places two objects side by side for comparison or contrast. For example, Milton juxtaposes God and Satan in "Paradise Lost."

Paradox: A paradox is a statement that appears self-contradictory but is actually true. One example of a paradox is when Socrates said, "I know one thing; that I know nothing." Seemingly, if Socrates knew nothing, he wouldn't know that he knew nothing. However, he is using figurative language not to say that he literally knows nothing, but that true wisdom begins with casting all presuppositions about the world aside.

Hyperbole: A hyperbole is an exaggeration. Ex: "I'm so tired I could sleep for centuries."

Allusion: An allusion is a reference to a character or event that happened in the past. T.S. Eliot's "The Waste Land" is a poem littered with allusions, including, "I will show you fear in a handful of dust," alluding to Genesis 3:19: "For you are dust, and to dust you shall return."

Pun: Puns are used in popular culture to invoke humor by exploiting the meanings of words. They can also be used in literature to give hints of meaning in unexpected places. In "Romeo and Juliet," Mercutio makes a pun after he is stabbed by Tybalt: "look for me tomorrow and you will find me a grave man."

Imagery: This is a collection of images given to the reader by the author. If a text is rich in imagery, it is easier for the reader to imagine themselves in the author's world. One example of a poem that relies on imagery is William Carlos Williams' "The Red Wheelbarrow":

so much depends
upon

a red wheel
barrow

glazed with rain
water

beside the white
chickens

The starkness of the imagery and the placement of the words in this poem bring to life the images of a purely simple world. Through its imagery, this poem tells a story in just sixteen words.

Symbolism: A symbol is used to represent an idea or belief system. For example, poets in Western civilization have been using the symbol of a rose for hundreds of years to represent love. In Japan, poets have used the firefly to symbolize passionate love, and sometimes even spirits of those who have died. Symbols can also express powerful political commentary and can be used in propaganda.

Irony: There are three types of irony: verbal, dramatic, and situational. **Verbal irony** is when a person states one thing and means the opposite. For example, a person is probably using irony when they say, "I can't wait to study for this exam next week." **Dramatic irony** occurs in a narrative and happens when the audience knows something that the characters do not. In the modern TV series Hannibal, the audience knows that Hannibal Lecter is a serial killer, but most of the main characters do not. This is dramatic irony. Finally, **situational irony** is when one expects something to happen, and the opposite occurs. For example, we can say that a fire station burning down would be an instance of situational irony.

Author's Intent

Authors may have many purposes for writing a specific text. They could be imparting information, entertaining their audience, expressing their own feelings, or trying to persuade their readers of a particular position. Authors' purposes are their reasons for writing something. A single author may have one overriding purpose for writing or multiple reasons. An author may explicitly state their intention in the text, or the reader may need to infer that intention. When readers can identify the author's purpose, they are better able to analyze information in the text. By knowing why the author wrote the text, readers can glean ideas for how to approach it. The following is a list of questions readers can ask in order to discern an author's purpose for writing a text:

- Does the title of the text give you any clues about its purpose?
- Was the purpose of the text to give information to readers?
- Did the author want to describe an event, issue, or individual?
- Was it written to express emotions and thoughts?
- Did the author want to convince readers to consider a particular issue?
- Do you think the author's primary purpose was to entertain?
- Why do you think the author wrote this text from a certain point of view?
- What is your response to the text as a reader?
- Did the author state their purpose for writing it?

Rather than simply consuming the text, readers should attempt to interpret the information being presented. Being able to identify an author's purpose efficiently improves reading comprehension, develops critical thinking, and makes students more likely to consider issues in depth before accepting writer viewpoints. Authors of fiction frequently write to entertain readers. Another purpose for writing fiction is making a political statement; for example, Jonathan Swift wrote "A Modest Proposal" (1729) as a political satire. Another purpose for writing fiction as well as nonfiction is to persuade readers to take some action or further a particular cause. Fiction authors and poets both frequently write to evoke certain moods; for example, Edgar Allan Poe wrote novels, short stories, and poems that evoke moods of gloom, guilt, terror, and dread. Another purpose of poets is evoking certain emotions: love is popular, as in Shakespeare's sonnets and numerous others. In "The Waste Land" (1922), T.S. Eliot evokes society's alienation, disaffection, sterility, and fragmentation.

Authors seldom directly state their purposes in texts. Some students may be confronted with nonfiction texts such as biographies, histories, magazine and newspaper articles, and instruction manuals, among others. To identify the purpose in nonfiction texts, students can ask the following questions:

- Is the author trying to teach something?
- Is the author trying to persuade the reader?
- Is the author imparting factual information only?
- Is this a reliable source?
- Does the author have some kind of hidden agenda?

To apply author purpose in nonfictional passages, students can also analyze sentence structure, word choice, and transitions to answer the aforementioned questions and to make inferences. For example, authors wanting to convince readers to view a topic negatively often choose words with negative connotations.

Narrative Writing

Narrative writing tells a story. The most prominent type of narrative writing is the fictional novel. Here are some examples:

- Mark Twain's *The Adventures of Tom Sawyer and The Adventures of Huckleberry Finn*
- Victor Hugo's *Les Misérables*
- Charles Dickens' *Great Expectations, David Copperfield*, and *A Tale of Two Cities*
- Jane Austen's *Northanger Abbey, Pride and Prejudice, Sense and Sensibility*, and *Emma*
- Toni Morrison's *Beloved, The Bluest Eye*, and *Song of Solomon*
- Gabriel García Márquez's *One Hundred Years of Solitude* and *Love in the Time of Cholera*

Nonfiction works can also appear in narrative form. For example, some authors choose a narrative style to convey factual information about a topic, such as a specific animal, country, geographic region, and scientific or natural phenomenon.

Narrative writing tells a story, and the one telling the story is called the narrator. The narrator may be a fictional character telling the story from their own viewpoint. This narrator uses the **first person** (*I, me, my, mine* and *we, us, our,* and *ours*). The narrator may also be the author; for example, when Louisa May Alcott writes "Dear reader" in *Little Women*, she (the author) addresses us as readers. In this case, the novel is typically told in **third person**, referring to the characters as *he, she, they*, or *them*. Another more common technique is the **omniscient narrator**; in other words, the story is told by an unidentified individual who sees and knows everything about the events and characters—not only their externalized actions, but also their internalized feelings and thoughts. **Second person** narration, which addresses readers as *you* throughout the text, is more uncommon than the first and third person options.

Expository Writing

Expository writing is also known as informational writing. Its purpose is not to tell a story as in narrative writing, to paint a picture as in descriptive writing, or to persuade readers to agree with something as in argumentative writing. Rather, its point is to communicate information to the reader. As such, the point of view of the author will necessarily be more objective. **Expository writing** does not appeal to emotions or reason, nor does it use subjective descriptions to sway the reader's opinion or thinking; rather, expository writing simply provides facts, evidence, observations, and objective descriptions of the subject matter. Some examples of expository writing include research reports, journal articles, books about history, academic textbooks, essays, how-to articles, user instruction manuals, news articles, and other factual journalistic reports.

Technical Writing

Technical writing is similar to expository writing because it provides factual and objective information. Indeed, it may even be considered a subcategory of expository writing. However, technical writing differs from expository writing in two ways: (1) it is specific to a particular field, discipline, or subject, and (2) it uses technical terminology that belongs only to that area. Writing that uses technical terms is intended only for an audience familiar with those terms. An example of technical writing would be a manual on computer programming and use.

Persuasive Writing

Persuasive writing, or **argumentative writing**, attempts to convince the reader to agree with the author's position. Some writers may respond to other writers' arguments by making reference to those authors or texts and then disagreeing with them. However, another common technique is for the author to anticipate opposing viewpoints, both from other authors and from readers. The author brings up these opposing viewpoints, and then refutes them before they can even be raised, strengthening the author's argument. Writers persuade readers by appealing to the readers' reason and emotion, as well as to their own character and credibility. Aristotle called these appeals **logos**, **pathos**, and **ethos**, respectively.

Evaluating the Author's Point of View in a Given Text

When a writer tells a story using the first person, readers can identify this by the use of first-person pronouns, like *I, me, we, us,* etc. However, first-person narratives can be told by different people or from different points of view. For example, some authors write in the first person to tell the story from the main character's viewpoint, as Charles Dickens did in his novels *David Copperfield* and *Great Expectations.* Some authors write in the first person from the viewpoint of a fictional character in the story, but not necessarily the main character. For example, F. Scott Fitzgerald wrote *The Great Gatsby* as narrated by Nick Carraway, a character in the story, about the main characters, Jay Gatsby and Daisy Buchanan. Other authors write in the first person, but as the omniscient narrator—an often unnamed person who knows all of the characters' inner thoughts and feelings. Writing in first person as oneself is more common in nonfiction.

Third Person

The **third-person** narrative is probably the most prevalent voice used in fictional literature. While some authors tell stories from the point of view and in the voice of a fictional character using the first person, it is a more common practice to describe the actions, thoughts, and feelings of fictional characters in the third person using *he, him, she, her, they, them,* etc.

Although plot and character development are both necessary and possible when writing narrative from a first-person point of view, they are also more difficult, particularly for new writers and those who find it unnatural or uncomfortable to write from that perspective. Therefore, writing experts advise beginning writers to start out writing in the third person. A big advantage of third-person narration is that the writer can describe the thoughts, feelings, and motivations of every character in a story, which is not possible for the first-person narrator. Third-person narrative can impart information to readers that the characters do not know. On the other hand, beginning writers often regard using the third-person point of view as more difficult because they must write about the feelings and thoughts of every character, rather than only about those of the protagonist.

Second Person

Narrative texts written in the **second person** address someone else as "you." In novels and other fictional works, the second person is the narrative voice most seldom used. The primary reason for this is that it often reads in an awkward manner, which prevents readers from being drawn into the fictional world of the novel. The second person is more often used in informational text, especially in how-to manuals, guides, and other instructions.

First Person

First person uses pronouns such as *I, me, we, my, us,* and *our.* Some writers naturally find it easier to tell stories from their own points of view, so writing in the first person offers advantages for them. The first-person voice is better for interpreting the world from a single viewpoint, and for enabling reader immersion in one protagonist's experiences. However, others find it difficult to use the first-person narrative voice. Its disadvantages can include overlooking the emotions of characters, forgetting to include description, producing stilted writing, using too many sentence structures involving "I did....", and not devoting enough attention to the story's "here-and-now" immediacy.

Selecting Important Points

The **topic** of a text is the general subject matter. Text topics can usually be expressed in one word, or a few words at most. Additionally, readers should ask themselves what point the author is trying to make. This point is the **main idea** of the text, the one thing the author wants readers to know concerning the topic. Once the author has established the main idea, they will support the main idea by supporting details. **Supporting details** are evidence that support the main idea and include personal testimonies, examples, or statistics.

One analogy for these components and their relationships is that a text is like a well-designed house. The topic is the roof, covering all rooms. The main idea is the frame. The supporting details are the various rooms. To identify the topic of a text, readers can ask themselves what or who the author is writing about in the paragraph. To locate the main idea, readers can ask themselves what one idea the author wants readers to know about the topic. To identify supporting details, readers can put the main idea into question form and ask "what does the author use to prove or explain their main idea?"

Let's look at an example. An author is writing an essay about the Amazon rainforest and trying to convince the audience that more funding should go into protecting the area from deforestation. The author makes the argument stronger by including evidence of the benefits of the rainforest: it provides habitats to a variety of species, it provides much of the earth's oxygen which in turn cleans the atmosphere, and it is the home to medicinal plants that may be the answer to some of the world's deadliest diseases.

Here is an outline of the essay looking at topic, main idea, and supporting details:

- **Topic:** Amazon rainforest
- **Main Idea:** The Amazon rainforest should receive more funding in order to protect it from deforestation.
- **Supporting Details:**
    - 1. It provides habitats to a variety of species
    - 2. It provides much of the earth's oxygen which in turn cleans the atmosphere
    - 3. It is home to medicinal plants that may be the answer to some of the world's deadliest diseases.

Notice that the topic of the essay is listed in a few key words: "Amazon rainforest." The main idea tells us what about the topic is important: that the topic should be funded in order to prevent deforestation. Finally, the supporting details are what the author relies on to convince the audience to act or to believe in the truth of the main idea.

Summarizing a Text

An important skill is the ability to read a complex text and then reduce its length and complexity by focusing on the key events and details. A **summary** is a shortened version of the original text, written by the reader in their own words. The summary should be shorter than the original text, and it must include the most critical points.

In order to effectively summarize a complex text, it's necessary to understand the original source and identify the major points covered. It may be helpful to outline the original text to get the big picture and avoid getting bogged down in the minor details. For example, a summary wouldn't include a statistic from the original source unless it was the major focus of the text. It is also important for readers to use their own words but still retain the original meaning of the passage. The key to a good summary is emphasizing the main idea without changing the focus of the original information.

Complex texts will likely be more difficult to summarize. Readers must evaluate all points from the original source, filter out the unnecessary details, and maintain only the essential ideas. The summary often mirrors the original text's organizational structure. For example, in a problem-solution text structure, the author typically presents readers with a problem and then develops solutions through the course of the text. An effective summary would likely retain this general structure, rephrasing the problem and then reporting the most useful or plausible solutions.

Paraphrasing is somewhat similar to summarizing. It calls for the reader to take a small part of the passage and list or describe its main points. Paraphrasing is more than rewording the original passage, though. As with summary, a paraphrase should be written in the reader's own words, while still retaining the meaning of the original source. The main difference between summarizing and paraphrasing is that a summary would be appropriate for a much larger text, while paraphrase might focus on just a few lines of text. Effective paraphrasing will indicate an understanding of the original source, yet still help the reader expand on their interpretation. A paraphrase should neither add new information nor remove essential facts that change the meaning of the source.

Comparing and Contrasting Themes from Print and Other Sources

The **theme** of a text is the central idea the author communicates. Whereas the topic of a passage of text may be concrete in nature, by contrast, the theme is always conceptual. For example, while the topic of Mark Twain's novel *The Adventures of Huckleberry Finn* might be described as something like the coming-of-age experiences of a poor, illiterate, functionally orphaned boy around and on the Mississippi River in 19th-century Missouri, one theme of the book might be that human beings are corrupted by society. Another might be that slavery and "civilized" society itself are hypocritical. Whereas the main idea in a text is the most important single point that the author wants to make, the theme is the concept or view around which the author centers the text.

Throughout time, humans have told stories with similar themes. Some themes are universal across time, space, and culture. These include themes of the individual as a hero, conflicts of the individual against nature, the individual against society, change vs. tradition, the circle of life, coming-of-age, and the complexities of love. Themes involving war and peace have featured prominently in diverse works, like Homer's *Iliad*, Tolstoy's *War and Peace* (1869), Stephen Crane's *The Red Badge of Courage* (1895), Hemingway's *A Farewell to Arms* (1929), and Margaret Mitchell's *Gone with the Wind* (1936). Another universal literary theme is that of the quest. These appear in folklore from countries and cultures worldwide, including the Gilgamesh Epic, Arthurian legend's Holy Grail quest, Virgil's *Aeneid*, Homer's

Odyssey, and the *Argonautica*. Cervantes' *Don Quixote* is a parody of chivalric quests. J.R.R. Tolkien's *The Lord of the Rings* trilogy (1954) also features a quest.

Similar themes across cultures often occur in countries that share a border or are otherwise geographically close together. For example, a folklore story of a rabbit in the moon using a mortar and pestle is shared among China, Japan, Korea, and Thailand—making medicine in China, making rice cakes in Japan and Korea, and hulling rice in Thailand. Another instance is when cultures are more distant geographically, but their languages are related. For example, East Turkestan's Uighurs and people in Turkey share tales of folk hero Effendi Nasreddin Hodja. Another instance, which may either be called cultural diffusion or simply reflect commonalities in the human imagination, involves shared themes among geographically and linguistically different cultures: both Cameroon's and Greece's folklore tell of centaurs; Cameroon, India, Malaysia, Thailand, and Japan, of mermaids; Brazil, Peru, China, Japan, Malaysia, Indonesia, and Cameroon, of underwater civilizations; and China, Japan, Thailand, Vietnam, Malaysia, Brazil, and Peru, of shape-shifters.

Two prevalent literary themes are love and friendship, which can end happily, sadly, or both. William Shakespeare's *Romeo and Juliet*, Emily Brontë's *Wuthering Heights*, Leo Tolstoy's *Anna Karenina,* and both *Pride and Prejudice* and *Sense and Sensibility* by Jane Austen are famous examples. Another theme recurring in popular literature is that of revenge, an old theme in dramatic literature, e.g. Elizabethan Thomas Kyd's *The Spanish Tragedy* and Thomas Middleton's *The Revenger's Tragedy*. Some more well-known instances include Shakespeare's tragedies *Hamlet* and *Macbeth,* Alexandre Dumas' *The Count of Monte Cristo,* John Grisham's *A Time to Kill*, and Stieg Larsson's *The Girl Who Kicked the Hornet's Nest*.

Themes are underlying meanings in literature. For example, if a story's main idea is a character succeeding against all odds, the theme is overcoming obstacles. If a story's main idea is one character wanting what another character has, the theme is jealousy. If a story's main idea is a character doing something they were afraid to do, the theme is courage. Themes differ from topics in that a topic is a subject matter; a theme is the author's opinion about it. For example, a work could have a topic of war and a theme that war is a curse. Authors present themes through characters' feelings, thoughts, experiences, dialogue, plot actions, and events. Themes function as "glue" holding other essential story elements together. They offer readers insights into characters' experiences, the author's philosophy, and how the world works.

Structure of a Text

Text structure is the way in which the author organizes and presents textual information so readers can follow and comprehend it. One kind of text structure is **sequence**. This means the author arranges the text in a logical order from beginning to middle to end. There are three types of sequences:

- **Chronological**: ordering events in time from earliest to latest
- **Spatial**: describing objects, people, or spaces according to their relationships to one another in space
- **Order of Importance**: addressing topics, characters, or ideas according to how important they are, from either least important to most important

Chronological sequence is the most common sequential text structure. Readers can identify sequential structure by looking for words that signal it, like *first, earlier, meanwhile, next, then, later, finally;* and specific times and dates the author includes as chronological references.

Problem-Solution Text Structure

The **problem-solution text structure** organizes textual information by presenting readers with a problem and then developing its solution throughout the course of the text. The author may present a variety of alternatives as possible solutions, eliminating each as they are found unsuccessful, or gradually leading up to the ultimate solution. For example, in fiction, an author might write a murder mystery novel and have the character(s) solve it through investigating various clues or character alibis until the killer is identified. In nonfiction, an author writing an essay or book on a real-world problem might discuss various alternatives and explain their disadvantages or why they would not work before identifying the best solution. For scientific research, an author reporting and discussing scientific experiment results would explain why various alternatives failed or succeeded.

Comparison-Contrast Text Structure

Comparison identifies similarities between two or more things. **Contrast** identifies differences between two or more things. Authors typically employ both to illustrate relationships between things by highlighting their commonalities and deviations. For example, a writer might compare Windows and Linux as operating systems, and contrast Linux as free and open-source vs. Windows as proprietary. When writing an essay, sometimes it is useful to create an image of the two objects or events you are comparing or contrasting. **Venn diagrams** are useful because they show the differences as well as the similarities between two things. Once you've seen the similarities and differences on paper, it might be helpful to create an outline of the essay with both comparison and contrast. Every outline will look different, because every two or more things will have a different number of comparisons and contrasts. Say you are trying to compare and contrast carrots with sweet potatoes.

Here is an example of a compare/contrast outline using those topics:

- Introduction: Talk about why you are comparing and contrasting carrots and sweet potatoes. Give the thesis statement.
- Body paragraph 1: Sweet potatoes and carrots are both root vegetables (similarity)
- Body paragraph 2: Sweet potatoes and carrots are both orange (similarity)
- Body paragraph 3: Sweet potatoes and carrots have different nutritional components (difference)
- Conclusion: Restate the purpose of your comparison/contrast essay.

Of course, if there is only one similarity between your topics and two differences, you will want to rearrange your outline. Always tailor your essay to what works best with your topic.

Descriptive Text Structure

Description can be both a type of text structure and a type of text. Some texts are descriptive throughout entire books. For example, a book may describe the geography of a certain country, state, or region, or tell readers all about dolphins by describing many of their characteristics. Many other texts are not descriptive throughout, but use descriptive passages within the overall text. The following are a few examples of descriptive text:

- When the author describes a character in a novel
- When the author sets the scene for an event by describing the setting
- When a biographer describes the personality and behaviors of a real-life individual
- When a historian describes the details of a particular battle within a book about a specific war
- When a travel writer describes the climate, people, foods, and/or customs of a certain place

A hallmark of description is using **sensory details**, painting a vivid picture so readers can imagine it almost as if they were experiencing it personally.

Cause and Effect Text Structure

When using **cause and effect** to extrapolate meaning from text, readers must determine the cause when the author only communicates effects. For example, if a description of a child eating an ice cream cone includes details like beads of sweat forming on the child's face and the ice cream dripping down her hand faster than she can lick it off, the reader can infer or conclude it must be hot outside. A useful technique for making such decisions is wording them in "If...then" form, e.g. "If the child is perspiring and the ice cream melting, it may be a hot day." Cause and effect text structures explain why certain events or actions resulted in particular outcomes. For example, an author might describe America's historical large flocks of dodo birds, the fact that gunshots did not startle/frighten dodos, and that because dodos did not flee, settlers killed whole flocks in one hunting session, explaining how the dodo was hunted into extinction.

Recognizing Events in a Sequence

Sequence structure is the order of events in which a story or information is presented to the audience. Sometimes the text will be presented in chronological order, or sometimes it will be presented by displaying the most recent information first, then moving backwards in time. The sequence structure depends on the author, the context, and the audience. The structure of a text also depends on the genre in which the text is written. Is it literary fiction? Is it a magazine article? Is it instructions for how to complete a certain task? Different genres will have different purposes for switching up the sequence.

Narrative Structure

The structure presented in literary fiction, called **narrative structure**, is the foundation on which the text moves. The narrative structure comes from the plot and setting. The **plot** is the sequence of events in the narrative that moves the text forward through cause and effect. The **setting** is the place or time period in which the story takes place. Narrative structure has two main categories: **linear** and **nonlinear**.

Linear Narrative

A narrative is **linear** when it is told in chronological order. Traditional linear narratives will follow the plot diagram below depicting the narrative arc. The **narrative arc** consists of the exposition, conflict, rising action, climax, falling action, and resolution.

- **Exposition**: The exposition is in the beginning of a narrative and introduces the characters, setting, and background information of the story. The exposition provides the context for the upcoming narrative. Exposition literally means "a showing forth" in Latin.

- **Conflict**: In a traditional narrative, the conflict appears toward the beginning of the story after the audience becomes familiar with the characters and setting. The conflict is a single instance between characters, nature, or the self, in which the central character is forced to make a decision or move forward with some kind of action. The conflict presents something for the main character, or protagonist, to overcome.

- **Rising Action**: The rising action is the part of the story that leads into the climax. The rising action will develop the characters and plot while creating tension and suspense that eventually lead to the climax.

- **Climax**: The climax is the part of the story where the tension produced in the rising action comes to a culmination. The climax is the peak of the story. In a traditional structure, everything before the climax builds up to it, and everything after the climax falls from it. It is the height of the narrative, and it is usually either the most exciting part of the story or a turning point in the character's journey.
- **Falling Action**: The falling action happens as a result of the climax. Characters continue to develop, although there is a wrapping up of loose ends here. The falling action leads to the resolution.
- **Resolution**: The resolution is where the story comes to an end and usually leaves the reader with the satisfaction of knowing what happened within the story and why. However, stories do not always end in this fashion. Sometimes readers can be confused or frustrated at the end from lack of information or the absence of a happy ending.

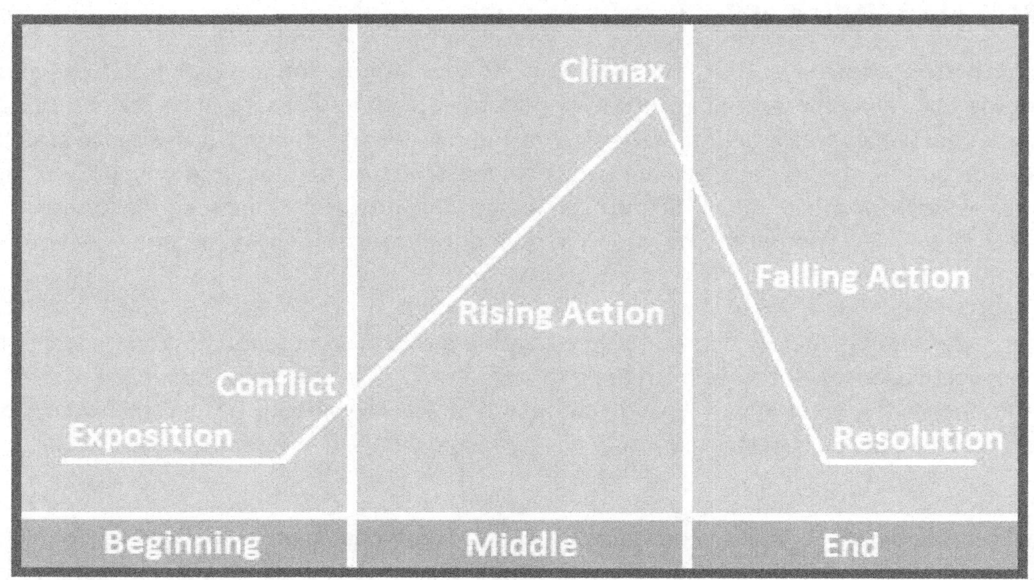

Nonlinear Narrative

A **nonlinear** narrative deviates from the traditional narrative because it does not always follow the traditional plot structure of the narrative arc. Nonlinear narratives may include structures that are disjointed, circular, or disruptive, in the sense that they do not follow chronological order. **In medias res** is an example of a nonlinear structure. *In medias res* is Latin for "in the middle of things," which is how many ancient texts, especially epic poems, began their story, such as Homer's *Iliad.* Instead of having a clear exposition with a full development of characters, they would begin right in the middle of the action.

Many modernist texts in the late nineteenth and early twentieth centuries experimented with disjointed narratives, moving away from traditional linear narratives. Disjointed narratives are depicted in novels like *Catch 22*, where the author, Joseph Heller, structures the narrative based on free association of ideas rather than chronology. Another nonlinear narrative can be seen in the novel *Wuthering Heights*, written by Emily Brontë; after the first chapter, the narrative progresses retrospectively instead of chronologically. There seem to be two narratives in *Wuthering Heights* working at the same time: a present narrative as well as a past narrative. Authors employ disrupting narratives for various reasons; some use it for the purpose of creating situational irony for the readers, while some use it to create a certain effect, such as excitement, discomfort, or fear.

Sequence Structure in Technical Documents

The purpose of technical documents, such as instruction manuals, cookbooks, or "user-friendly" documents, is to provide information to users as clearly and efficiently as possible. In order to do this, the sequence structure in technical documents should be as straightforward as possible. This usually involves some kind of chronological order or a direct sequence of events. For example, someone who is reading an instruction manual on how to set up their Smart TV wants directions in a clear, simple, straightforward manner that does not confuse them or leave them guessing about the proper sequence.

Sequence Structure in Informational Texts

The structure of informational texts depends on the specific genre. For example, a newspaper article may start by stating an exciting event that happened, then talk about that event in chronological order. Many informational texts also use **cause and effect structure**, which describes an event and then identifies reasons for why that event occurred. Some essays may write about their subjects by way of **comparison and contrast**, which is a structure that compares two things or contrasts them to highlight their differences. Other documents, such as proposals, will have a **problem to solution structure**, where the document highlights some kind of problem and then offers a solution. Finally, some informational texts are written with lush details and description in order to captivate the audience, allowing them to visualize the information presented to them. This type of structure is known as **descriptive**.

Meanings of Words, Sentences, and Entire Texts

When readers encounter an unfamiliar word in a text, they can use the surrounding context—the overall subject matter, specific chapter/section topic, and especially the immediate sentence context. Among others, one category of context clues is grammar. For example, the position of a word in a sentence and its relationship to the other words can help the reader establish whether the unfamiliar word is a verb, a noun, an adjective, an adverb, etc. This narrows down the possible meanings of the word to one part of speech. However, this may be insufficient. In the sentence, "Many birds migrate twice yearly," the reader can determine that the italicized word is a verb. While it probably does not mean eat or drink (because birds would need to do those actions more than twice each year), it could mean travel, mate, lay eggs, hatch, molt, etc.

Some words can have a number of different meanings depending on how they are used. For example, the word *fly* has a different meaning in each of the following sentences:

- "His trousers have a fly on them."
- "He swatted the fly on his trousers."
- "Those are some fly trousers."
- "They went fly fishing."
- "She hates to fly."
- "If humans were meant to fly, they would have wings."

As strategies, readers can try substituting a familiar word for an unfamiliar one and see whether it makes sense in the sentence. They can also identify other words in a sentence, offering clues to an unfamiliar word's meaning.

Denotation and Connotation

Denotation, a word's explicit definition, is often set in comparison to **connotation**, the emotional, cultural, social, or personal implications associated with a word. Denotation is more of an objective definition, whereas connotation can be more subjective, although many connotative meanings of words are similar for certain cultures. The denotative meanings of words are usually based on facts, and the connotative meanings of words are usually based on emotion.

Here are some examples of words and their denotative and connotative meanings in Western culture:

Word	Denotative Meaning	Connotative Meaning
Home	A permanent place where one lives, usually as a member of a family.	A place of warmth; a place of familiarity; comforting; a place of safety and security. "Home" usually has a positive connotation.
Snake	A long reptile with no limbs and strong jaws that moves along the ground; some snakes have a poisonous bite.	An evil omen; a slithery creature (human or nonhuman) that is deceitful or unwelcome. "Snake" usually has a negative connotation.
Winter	A season of the year that is the coldest, usually from December to February in the northern hemisphere and from June to August in the southern hemisphere.	Circle of life, especially that of death and dying; cold or icy; dark and gloomy; hibernation, sleep, or rest. Winter can have a negative connotation, although many who have access to heat may enjoy the snowy season from their homes.

Context Clues

Readers can often figure out what unfamiliar words mean without interrupting their reading to look them up in dictionaries by examining **context clues**. Context includes the other words or sentences in a passage. One common context clue is the root word and any affixes (prefixes/suffixes). Another common context clue is a synonym or definition included in the sentence. Sometimes both exist in the same sentence. Here's an example:

> Scientists who study birds are *ornithologists*.

Many readers may not know the word *ornithologist*. However, the example contains a definition (scientists who study birds). The reader may also have the ability to analyze the suffix (*-logy*, meaning the study of) and root (*ornitho-*, meaning bird).

Another common context clue is a sentence that shows differences. Here's an example:

> Birds *incubate* their eggs outside of their bodies, unlike mammals.

Some readers may be unfamiliar with the word *incubate*. However, since the sentence includes the phrase "unlike mammals," the reader can infer that incubate relates to an aspect of prenatal development that mammals and birds do not have in common, such as the way they keep the embryo at a temperature suitable for development.

In addition to analyzing the etymology of a word's root and affixes and extrapolating word meaning from sentences that contrast an unknown word with an antonym, readers can also determine word meanings from sentence context clues based on logic. Here's an example:

> Birds are always looking out for predators that could attack their young.

The reader who is unfamiliar with the word *predator* could determine from the context of the sentence that predators usually prey upon baby birds and possibly other young animals. Readers might also use the context clue of etymology here, as *predator* and *prey* have the same root.

Analyzing Word Parts

Readers can learn some **etymologies**, or origins, of words and their parts, making it easier to break down new words into components and analyze their combined meanings. For example, the root word soph is Greek for "wise" or "knowledge." Knowing this informs the meanings of English words including *sophomore, sophisticated,* and *philosophy.* Those who also know that phil is Greek for "love" will realize that philosophy means "love of knowledge." They can then extend this knowledge of *phil* to understand *philanthropist* (one who loves people), *bibliophile* (book lover), *philharmonic* (loving harmony), *hydrophilic* (water-loving), and so on. In addition, *phob* derives from the Greek *phobos,* meaning "fear." Words with this root indicate fear of various things *acrophobia* (fear of heights), *arachnophobia* (fear of spiders), *claustrophobia* (fear of enclosed spaces), *ergophobia* (fear of work), and *hydrophobia* (fear of water), among others.

Some English word origins from other languages, like ancient Greek, are found in large numbers and varieties of English words. An advantage of the shared ancestry of these words is that once readers recognize the meanings of some Greek words or word roots, they can determine or at least get an idea of what many different English words mean. As an example, the Greek word *métron* means to measure, a measure, or something used to measure; the English word meter derives from it. Knowing this informs many other English words, including *altimeter, barometer, diameter, hexameter, isometric,* and *metric.* While readers must know the meanings of the other parts of these words to decipher their meaning fully, they already have an idea that they are all related in some way to measures or measuring.

While all English words ultimately derive from a proto-language known as Indo-European, many of them historically came into the developing English vocabulary later, from sources like the ancient Greeks' language, the Latin used throughout Europe and much of the Middle East during the reign of the Roman Empire, and the Anglo-Saxon languages used by England's early tribes. In addition to classic revivals and native foundations, by the Renaissance era other influences included French, German, Italian, and Spanish. Today we can often discern English word meanings by knowing common roots and affixes, particularly from Greek and Latin.

The following is a list of common prefixes and their meanings:

Prefix	Definition	Examples
a-	without	atheist, agnostic
ad-	to, toward	advance
ante-	before	antecedent, antedate
anti-	opposing	antipathy, antidote
auto-	self	autonomy, autobiography
bene-	well, good	benefit, benefactor
bi-	two	bisect, biennial
bio-	life	biology, biosphere
chron-	time	chronometer, synchronize
circum-	around	circumspect, circumference
com-	with, together	commotion, complicate
contra-	against, opposing	contradict, contravene
cred-	belief, trust	credible, credit
de-	from	depart
dem-	people	demographics, democracy
dis-	away, off, down, not	dissent, disappear
equi-	equal, equally	equivalent
ex-	from, out of	extract
for-	away, off, from	forget, forswear
fore-	before, previous	foretell, forefathers
homo-	same, equal	homogenized
hyper-	excessive, over	hypercritical, hypertension
in-	in, into	intrude, invade
inter-	among, between	intercede, interrupt
mal-	bad, poorly, not	malfunction
micr(o)-	small	microbe, microscope
mis-	bad, poorly, not	misspell, misfire
mono-	one, single	monogamy, monologue
mor-	die, death	mortality, mortuary
neo-	new	neolithic, neoconservative
non-	not	nonentity, nonsense
omni-	all, everywhere	omniscient
over-	above	overbearing
pan-	all, entire	panorama, pandemonium
para-	beside, beyond	parallel, paradox
phil-	love, affection	philosophy, philanthropic
poly-	many	polymorphous, polygamous
pre-	before, previous	prevent, preclude
prim-	first, early	primitive, primary
pro-	forward, in place of	propel, pronoun
re-	back, backward, again	revoke, recur

Prefix	Definition	Examples
sub-	under, beneath	subjugate, substitute
super-	above, extra	supersede, supernumerary
trans-	across, beyond, over	transact, transport
ultra-	beyond, excessively	ultramodern, ultrasonic, ultraviolet
un-	not, reverse of	unhappy, unlock
vis-	to see	visage, visible

The following is a list of common suffixes and their meanings:

Suffix	Definition	Examples
-able	likely, able to	capable, tolerable
-ance	act, condition	acceptance, vigilance
-ard	one that does excessively	drunkard, wizard
-ation	action, state	occupation, starvation
-cy	state, condition	accuracy, captaincy
-er	one who does	teacher
-esce	become, grow, continue	convalesce, acquiesce
-esque	in the style of, like	picturesque, grotesque
-ess	feminine	waitress, lioness
-ful	full of, marked by	thankful, zestful
-ible	able, fit	edible, possible, divisible
-ion	action, result, state	union, fusion
-ish	suggesting, like	churlish, childish
-ism	act, manner, doctrine	barbarism, socialism
-ist	doer, believer	monopolist, socialist
-ition	action, result, state,	sedition, expedition
-ity	quality, condition	acidity, civility
-ize	cause to be, treat with	sterilize, mechanize, criticize
-less	lacking, without	hopeless, countless
-like	like, similar	childlike, dreamlike
-ly	like, of the nature of	friendly, positively
-ment	means, result, action	refreshment, disappointment
-ness	quality, state	greatness, tallness
-or	doer, office, action	juror, elevator, honor
-ous	marked by, given to	religious, riotous
-some	apt to, showing	tiresome, lonesome
-th	act, state, quality	warmth, width
-ty	quality, state	enmity, activity

Practice Questions

Section 1

Reading Comprehension

The next four questions are based on the following passage from The Life, Crime, and Capture of John Wilkes Booth *by George Alfred Townsend:*

> The box in which the President sat consisted of two boxes turned into one, the middle partition being removed, as on all occasions when a state party visited the theater. The box was on a level with the dress circle; about twelve feet above the stage. There were two entrances—the door nearest to the wall having been closed and locked; the door nearest the balustrades of the dress circle, and at right angles with it, being open and left open, after the visitors had entered. The interior was carpeted, lined with crimson paper, and furnished with a sofa covered with crimson velvet, three arm chairs similarly covered, and six cane-bottomed chairs. Festoons of flags hung before the front of the box against a background of lace.
>
> President Lincoln took one of the arm-chairs and seated himself in the front of the box, in the angle nearest the audience, where, partially screened from observation, he had the best view of what was transpiring on the stage. Mrs. Lincoln sat next to him, and Miss Harris in the opposite angle nearest the stage. Major Rathbone sat just behind Mrs. Lincoln and Miss Harris. These four were the only persons in the box.
>
> The play proceeded, although "Our American Cousin," without Mr. Sothern, has, since that gentleman's departure from this country, been justly esteemed a very dull affair. The audience at Ford's, including Mrs. Lincoln, seemed to enjoy it very much. The worthy wife of the President leaned forward, her hand upon her husband's knee, watching every scene in the drama with amused attention. Even across the President's face at intervals swept a smile, robbing it of its habitual sadness.
>
> About the beginning of the second act, the mare, standing in the stable in the rear of the theater, was disturbed in the midst of her meal by the entrance of the young man who had quitted her in the afternoon. It is presumed that she was saddled and bridled with exquisite care.
>
> Having completed these preparations, Mr. Booth entered the theater by the stage door; summoned one of the scene shifters, Mr. John Spangler, emerged through the same door with that individual, leaving the door open, and left the mare in his hands to be held until he (Booth) should return. Booth who was even more fashionably and richly dressed than usual, walked thence around to the front of the theater, and went in. Ascending to the dress circle, he stood for a little time gazing around upon the audience and occasionally upon the stage in his usual graceful manner. He was subsequently observed by Mr. Ford, the proprietor of the theater, to be slowly elbowing his way through the crowd that packed the rear of the dress circle toward the right side, at the extremity of which was the box where Mr. and Mrs. Lincoln and their companions were seated. Mr. Ford casually noticed this as a slightly extraordinary symptom of interest on the part of an actor so familiar with the routine of the theater and the play.

1. Which of the following best describes the author's attitude toward the events leading up to the assassination of President Lincoln?
   a. Excitement due to the setting and its people.
   b. Sadness due to the death of a beloved president.
   c. Anger because of the impending violence.
   d. Neutrality due to the style of the report.
   e. Apprehension due to the crowd and their ignorance.

2. What does the author mean by the last sentence in the passage?
   a. Mr. Ford was suspicious of Booth and assumed he was making his way to Mr. Lincoln's box.
   b. Mr. Ford assumed Booth's movement throughout the theater was due to being familiar with the theater.
   c. Mr. Ford thought that Booth was making his way to the theater lounge to find his companions.
   d. Mr. Ford thought that Booth was elbowing his way to the dressing room to get ready for the play.
   e. Mr. Ford thought that Booth was coming down with an illness due to the strange symptoms he displayed.

3. Given the author's description of the play "Our American Cousin," which one of the following is most analogous to Mr. Sothern's departure from the theater?
   a. A ballet dancer who leaves the New York City Ballet just before they go on to their final performance.
   b. A basketball player leaves an NBA team and the next year they make it to the championship but lose.
   c. A lead singer leaves their band to begin a solo career, and the band drops in sales by 50 percent on their next album.
   d. A movie actor who dies in the middle of making a movie and the movie is made anyway by actors who resemble the deceased.
   e. A professor who switches to the top-rated university for their department only to find the university they left behind has surpassed his new department's rating.

4. Based on the organizational structure of the passage, which of the following texts most closely relates?
   a. A chronological account in a fiction novel of a woman and a man meeting for the first time.
   b. A cause-and-effect text ruminating on the causes of global warming.
   c. An autobiography that begins with the subject's death and culminates to his birth.
   d. A text focusing on finding a solution to the problem of the Higgs boson particle.
   e. A compare and contrast essay on the political stances of former US presidents.

The next three questions are based on the following passage from "Free Speech in War Time" by James Parker Hall, written in 1921, published in Columbia Law Review, *Vol. 21 No. 6:*

> In approaching this problem of interpretation, we may first put out of consideration certain obvious limitations upon the generality of all guaranties of free speech. An occasional unthinking malcontent may urge that the only meaning not fraught with danger to liberty is the literal one that no utterance may be forbidden, no matter what its intent or result; but in fact it is nowhere seriously argued by anyone whose opinion is entitled to respect that direct and intentional incitations to crime may not be forbidden by the state. If a state may properly forbid murder or robbery or treason, it may also punish those who induce or counsel the commission of such crimes. Any other view makes a mockery of the state's power to declare and punish offences. And what the state may do to prevent the incitement of serious crimes which are universally

condemned, it may also do to prevent the incitement of lesser crimes, or of those in regard to the bad tendency of which public opinion is divided. That is, if the state may punish John for burning straw in an alley, it may also constitutionally punish Frank for inciting John to do it, though Frank did so by speech or writing. And if, in 1857, the United States could punish John for helping a fugitive slave to escape, it could also punish Frank for inducing John to do this, even though a large section of public opinion might applaud John and condemn the Fugitive Slave Law.

5. The author uses the examples in the last lines in order to do what?

a. To compare different types of crimes to see by which one the principle of freedom of speech would become objectionable

b. To demonstrate that anyone who incites a crime, despite the severity or magnitude of the crime, should be held accountable for that crime in some degree

c. To prove that the definition of "freedom of speech" is altered depending on what kind of crime is being committed

d. To show that some crimes are in the best interest of a nation and should not be punishable if they are proven to prevent harm to others

e. To suggest that the crimes mentioned should be reopened in order to punish those who incited the crimes

6. Which of the following, if true, would most seriously undermine the claim proposed by the author in that if the state can punish a crime then it can punish the incitement of that crime?

a. The idea that human beings are able and likely to change their mind between the utterance and execution of an event that may harm others

b. The idea that human beings will always choose what they think is right based on their cultural upbringing

c. The idea that the limitation of free speech by the government during wartime will protect the country from any group that causes a threat to that country's freedom

d. The idea that those who support freedom of speech probably have intentions of subverting the government

e. The idea that if a man encourages a woman to commit a crime and she succeeds, the man is just as guilty as the woman

7. Which of the following words, if substituted for the word *malcontent*, would LEAST change the meaning of the sentence?

a. Regimen

b. Cacophony

c. Anecdote

d. Residual

e. Grievance

The next two questions are from Rhetoric and Poetry in the Renaissance: A Study of Rhetorical Terms in English Renaissance Literary Criticism *by D.L. Clark:*

To the Greeks and Romans rhetoric meant the theory of oratory. As a pedagogical mechanism it endeavored to teach students to persuade an audience. The content of rhetoric included all that the ancients had learned to be of value in persuasive public speech. It taught how to work up a case by drawing valid inferences from sound evidence, how to organize this material in the most persuasive order, how to compose in clear and harmonious sentences. Thus to the Greeks and

Romans rhetoric was defined by its function of discovering means to persuasion and was taught in the schools as something that every free-born man could and should learn.

In both these respects the ancients felt that poetics, the theory of poetry, was different from rhetoric. As the critical theorists believed that the poets were inspired, they endeavored less to teach men to be poets than to point out the excellences which the poets had attained. Although these critics generally, with the exceptions of Aristotle and Eratosthenes, believed the greatest value of poetry to be in the teaching of morality, no one of them endeavored to define poetry, as they did rhetoric, by its purpose. To Aristotle, and centuries later to Plutarch, the distinguishing mark of poetry was imitation. Not until the renaissance did critics define poetry as an art of imitation endeavoring to inculcate morality . . .

The same essential difference between classical rhetoric and poetics appears in the content of classical poetics. Whereas classical rhetoric deals with speeches which might be delivered to convict or acquit a defendant in the law court, or to secure a certain action by the deliberative assembly, or to adorn an occasion, classical poetic deals with lyric, epic, and drama. It is a commonplace that classical literary critics paid little attention to the lyric. It is less frequently realized that they devoted almost as little space to discussion of metrics. By far the greater bulk of classical treatises on poetics is devoted to characterization and to the technique of plot construction, involving as it does narrative and dramatic unity and movement as distinct from logical unity and movement.

8. What does the author say about one way in which the purpose of poetry changed for later philosophers?

a. The author says that at first, poetry was not defined by its purpose but was valued for its ability to be used to teach morality. Later, some philosophers would define poetry by its ability to instill morality. Finally, during the renaissance, poetry was believed to be an imitative art, but was not necessarily believed to instill morality in its readers.
b. The author says that the classical understanding of poetry dealt with its ability to be used to teach morality. Later, philosophers would define poetry by its ability to imitate life. Finally, during the renaissance, poetry was believed to be an imitative art that instilled morality in its readers.
c. The author says that at first, poetry was thought to be an imitation of reality, then later philosophers valued poetry more for its ability to instill morality.
d. The author says that the classical understanding of poetry was that it dealt with the search for truth through its content; later, the purpose of poetry would be through its entertainment.
e. The author says that the initial understanding of the purpose of poetry was through its entertainment. Then, as poetry evolved into a more religious era, the renaissance, it was valued for its ability to instill morality through its teaching.

9. What does the author of the passage say about classical literary critics in relation to poetics?
    a. That rhetoric was more valued than poetry because rhetoric had a definitive purpose to persuade an audience, and poetry's wavering purpose made it harder for critics to teach.
    b. That although most poetry was written as lyric, epic, or drama, the critics were most focused on the techniques of lyric and epic and their performance of musicality and structure.
    c. That although most poetry was written as lyric, epic, or drama, the critics were most focused on the techniques of the epic and drama and their performance of structure and character.
    d. That the study of poetics was more pleasurable than the study of rhetoric due to its ability to assuage its audience, and the critics therefore focused on what poets did to create that effect.
    e. That since poetics was made by the elite in Greek and Roman society, literary critics resented poetics for its obsession on material things and its superfluous linguistics.

Sentence Equivalence

Select the two answer choices that can complete the sentence and create sentences that have complementary meaning.

10. Although Courtney thought their realtor was quite moody while showing them houses that day, Chris found her to be quite ________.
    a. Belligerent
    b. Callous
    c. Amicable
    d. Genial
    e. Clandestine
    f. Destitute

11. The publishers purposely kept a(n) ________ mood surrounding the novel; they knew that its secrecy would enchant its potential readers.
    a. Demonstrative
    b. Enigmatic
    c. Apathetic
    d. Buoyant
    e. Reticent
    f. Convivial

12. She seemed to ________ in him a new sensation, probably what others would consider a taste of first love.
    a. Evoke
    b. Galvanize
    c. Placate
    d. Quell
    e. Raze
    f. Relinquish

13. Before they went in to look at the *Mona Lisa*, there was a certain kind of _________ that pervaded the crowd and made the onlookers anticipate the famous artwork even more.

a. Flippancy
b. Impudence
c. Reverence
d. Indolence
e. Tenacity
f. Deference

14. Their plan was to _________ the existing train station so that they could make room for a brand new baseball stadium.

a. Accost
b. Elicit
c. Reproach
d. Decimate
e. Eradicate
f. Annex

15. The meeting was about building a healthy _________ with coworkers and learning how to cope with stressors in the workplace.

a. Propinquity
b. Rancor
c. Rapport
d. Fortitude
e. Impasse
f. Hubris

16. The _________ nature of the phone conversation enraged the customer; treating someone like a child, he thought, was bad customer service and a negative reflection on the company as a whole.

a. Hackneyed
b. Supercilious
c. Innate
d. Prudent
e. Subtle
f. Patronizing

17. Kasey could hardly believe her eyes when she saw the end result of her brother's wedding: the _________ venue, the decorations, and the expensive flowers.

a. Frugal
b. Hardy
c. Inane
d. Opulent
e. Lucrative
f. Palatial

Text Completion

Select the best word from the corresponding column of choices that most clearly completes the passage:

18. Then, in the 1960s, the Sun Belt went from being less populated and impoverished to more popular and (i) ____________________. Today, baby boomers (ii) ________________ to the Sun Belt to retire, while a younger influx of people move here for (iii) ______________ opportunities.

Blank (i)	Blank (ii)	Blank (iii)
a. Destitute	d. Migrate	g. Puerile
b. Affluent	e. Exhort	h. Garrulous
c. Hapless	f. Rescind	i. Lucrative

19. Although she loved horseback riding, her (i) ________________ skills weren't enough to get her a scholarship into Auburn University, much to her (ii) __________________ .

Blank (i)	Blank (ii)
a. Equestrian	d. Blandishment
b. Equanimity	e. Mendaciousness
c. Flagrant	f. Consternation

20. Her (i) ____________ towards the outdoors would be the basis from which her career in Archaeology would stem.

Blank (i)
a. Tirade
b. Requisition
c. Proclivity
d. Negligence
e. Irreverence

21. His (i) _________________ decision to change out the night watchman sparked a (ii) _________________ of events that would lead to the sinking of the Titanic.

Blank (i)	Blank (ii)
a. Impetuous	d. Fetter
b. Ambivalent	e. Concatenation
c. Fortuitous	f. Turpitude

22. The volcanologist was (i) _________________ at finding out why certain volcanoes are dormant and others erupt, and discerning volcanic activity in past centuries.

Blank (i)
a. Banal
b. Adept
c. Divisive
d. Indignant
e. Precarious

23. The offer to the primary candidate was (i) ________________ after they found out that he was not actually qualified for the position. The otherwise (ii) ________________ hiring manager was horrified she could have let such an error occur.

Blank (i)	Blank (ii)
a. Sage	d. Punctilious
b. Benevolent	e. Bombastic
c. Rescinded	f. Disputatious

24. The play was (i) ____________ in a way that made outsiders uncomfortable; it seemed every other scene was a lesson of some sort.

Blank (i)
a. Elusive
b. Indulgent
c. Morose
d. Tenuous
e. Didactic

25. In the film, one of the characters was a(n) (i) ______________ figure who represented a long line of villains. The main character did his best to (ii) ________________ the terrifying figure, but the villain caught him in the end.

Blank (i)	Blank (ii)
a. Archetypal	d. Coalesce
b. Amenable	e. Circumvent
c. Antediluvian	f. Encompass

Section 2

The next four questions are based on the following passage from The Story of Germ Life *by Herbert William Conn:*

> The first and most universal change effected in milk is its souring. So universal is this phenomenon that it is generally regarded as an inevitable change which can not be avoided, and, as already pointed out, has in the past been regarded as a normal property of milk. To-day, however, the phenomenon is well understood. It is due to the action of certain of the milk bacteria upon the milk sugar which converts it into lactic acid, and this acid gives the sour taste and curdles the milk. After this acid is produced in small quantity its presence proves deleterious to the growth of the bacteria, and further bacterial growth is checked. After souring, therefore, the milk for some time does not ordinarily undergo any further changes.
>
> Milk souring has been commonly regarded as a single phenomenon, alike in all cases. When it was first studied by bacteriologists it was thought to be due in all cases to a single species of micro-organism which was discovered to be commonly present and named *Bacillus acidi lactici.* This bacterium has certainly the power of souring milk rapidly, and is found to be very common in dairies in Europe. As soon as bacteriologists turned their attention more closely to the subject it

was found that the spontaneous souring of milk was not always caused by the same species of bacterium. Instead of finding this *Bacillus acidi lactici* always present, they found that quite a number of different species of bacteria have the power of souring milk, and are found in different specimens of soured milk. The number of species of bacteria which have been found to sour milk has increased until something over a hundred are known to have this power. These different species do not affect the milk in the same way. All produce some acid, but they differ in the kind and the amount of acid, and especially in the other changes which are effected at the same time that the milk is soured, so that the resulting soured milk is quite variable. In spite of this variety, however, the most recent work tends to show that the majority of cases of spontaneous souring of milk are produced by bacteria which, though somewhat variable, probably constitute a single species, and are identical with the *Bacillus acidi lactici*. This species, found common in the dairies of Europe, according to recent investigations occurs in this country as well. We may say, then, that while there are many species of bacteria infesting the dairy which can sour the milk, there is one which is more common and more universally found than others, and this is the ordinary cause of milk souring.

When we study more carefully the effect upon the milk of the different species of bacteria found in the dairy, we find that there is a great variety of changes which they produce when they are allowed to grow in milk. The dairyman experiences many troubles with his milk. It sometimes curdles without becoming acid. Sometimes it becomes bitter, or acquires an unpleasant "tainted" taste, or, again, a "soapy" taste. Occasionally a dairyman finds his milk becoming slimy, instead of souring and curdling in the normal fashion. At such times, after a number of hours, the milk becomes so slimy that it can be drawn into long threads. Such an infection proves very troublesome, for many a time it persists in spite of all attempts made to remedy it. Again, in other cases the milk will turn blue, acquiring about the time it becomes sour a beautiful sky-blue colour. Or it may become red, or occasionally yellow. All of these troubles the dairyman owes to the presence in his milk of unusual species of bacteria which grow there abundantly.

1. The word *deleterious* in the first paragraph can be best interpreted as referring to which one of the following?
   a. Amicable
   b. Smoldering
   c. Luminous
   d. Ruinous
   e. Virtuous

2. Which of the following best explains how the passage is organized?
   a. The author begins by presenting the effects of a phenomenon, then explains the process of this phenomenon, and then ends by giving the history of the study of this phenomenon.
   b. The author begins by explaining a process or phenomenon, then gives the history of the study of this phenomenon, then ends by presenting the effects of this phenomenon.
   c. The author begins by giving the history of the study of a certain phenomenon, then explains the process of this phenomenon, then ends by presenting the effects of this phenomenon.
   d. The author begins by giving a broad definition of a subject, then presents more specific cases of the subject, then ends by contrasting two different viewpoints on the subject.
   e. The author begins by contrasting two different viewpoints, then gives a short explanation of a subject, then ends by summarizing what was previously stated in the passage.

3. What is the primary purpose of the passage?
   a. To inform the reader of the phenomenon, investigation, and consequences of milk souring.
   b. To persuade the reader that milk souring is due to *Bacillus acidi lactici*, found commonly in the dairies of Europe.
   c. To describe the accounts and findings of researchers studying the phenomenon of milk souring.
   d. To discount the former researchers' opinions on milk souring and bring light to new investigations.
   e. To narrate the story of one researcher who discovered the phenomenon of milk souring and its subsequent effects.

4. What does the author say about the ordinary cause of milk souring?
   a. Milk souring is caused mostly by a species of bacteria called *Bacillus acidi lactici*, although former research asserted that it was caused by a variety of bacteria.
   b. The ordinary cause of milk souring is unknown to current researchers, although former researchers thought it was due to a species of bacteria called *Bacillus acidi lactici*.
   c. Milk souring is caused mostly by a species of bacteria identical to that of *Bacillus acidi lactici*, although there are a variety of other bacteria that cause milk souring as well.
   d. The ordinary cause of milk souring will sometimes curdle without becoming acidic, though sometimes it will turn colors other than white, or have strange smells or tastes.
   e. The ordinary cause of milk souring is from bacteria with a strange, "soapy" smell, usually the color of sky blue.

The next question is based on the following passage from "Freedom of Speech in War Time" by Zechariah Chafee, Jr. written in 1919, published in Harvard Law Review *Vol. 32 No. 8):*

> The true boundary line of the First Amendment can be fixed only when Congress and the courts realize that the principle on which speech is classified as lawful or unlawful involves the balancing against each other of two very important social interests, in public safety and in the search for truth. Every reasonable attempt should be made to maintain both interests unimpaired, and the great interest in free speech should be sacrificed only when the interest in public safety is really imperiled, and not, as most men believe, when it is barely conceivable that it may be slightly affected. In war time, therefore, speech should be unrestricted by the censorship or by punishment, unless it is clearly liable to cause direct and dangerous interference with the conduct of the war.
>
> Thus our problem of locating the boundary line of free speech is solved. It is fixed close to the point where words will give rise to unlawful acts. We cannot define the right of free speech with the precision of the Rule against Perpetuities or the Rule in Shelley's Case, because it involves national policies which are much more flexible than private property, but we can establish a workable principle of classification in this method of balancing and this broad test of certain danger. There is a similar balancing in the determination of what is "due process of law." And we can with certitude declare that the First Amendment forbids the punishment of words merely for their injurious tendencies. The history of the Amendment and the political function of free speech corroborate each other and make this conclusion plain.

5. What is the primary purpose of the passage?
   a. To analyze the First Amendment in historical situations in order to make an analogy to the current war at hand in the nation
   b. To demonstrate that the boundaries set during wartime are different from that when the country is at peace, and that we should change our laws accordingly
   c. To offer the idea that during wartime, the principle of freedom of speech should be limited to that of even minor utterances in relation to a crime
   d. To claim the interpretation of freedom of speech is already evident in the First Amendment and to offer a clear perimeter of the principle during war time
   e. To assert that any limitation on freedom of speech is a violation of human rights and that the circumstances of war do not change this violation

The next two questions are based off the following passage:

> Rehabilitation rather than punitive justice is becoming much more popular in prisons around the world. Prisons in America, especially, where the recidivism rate is 67 percent, would benefit from mimicking prison tactics in Norway, which has a recidivism rate of only 20 percent. In Norway, the idea is that a rehabilitated prisoner is much less likely to offend than one harshly punished. Rehabilitation includes proper treatment for substance abuse, psychotherapy, healthcare and dental care, and education programs.

6. Which of the following best captures the author's purpose?
   a. To show the audience one of the effects of criminal rehabilitation by comparison
   b. To persuade the audience to donate to American prisons for education programs
   c. To convince the audience of the harsh conditions of American prisons
   d. To inform the audience of the incredibly lax system of Norwegian prisons
   e. To educate the audience on prison rehabilitation history and terminology

7. Which of the following describes the word *recidivism* as it is used in the passage?
   a. The lack of violence in the prison system.
   b. The opportunity of inmates to receive therapy in prison.
   c. The event of a prisoner escaping the compound.
   d. The likelihood of a convicted criminal to reoffend.
   e. The act of proving a statement to be false.

The next two questions are based off the following passage from Virginia Woolf's Mrs. Dalloway*:*

> What a lark! What a plunge! For so it had always seemed to her, when, with a little squeak of the hinges, which she could hear now, she had burst open the French windows and plunged at Bourton into the open air. How fresh, how calm, stiller than this of course, the air was in the early morning; like the flap of a wave; the kiss of a wave; chill and sharp and yet (for a girl of eighteen as she then was) solemn, feeling as she did, standing there at the open window, that something awful was about to happen; looking at the flowers, at the trees with the smoke winding off them and the rooks rising, falling; standing and looking until Peter Walsh said, "Musing among the vegetables?"—was that it?—"I prefer men to cauliflowers"—was that it? He must have said it at breakfast one morning when she had gone out on to the terrace—Peter Walsh. He would be back from India one of these days, June or July, she forgot which, for his letters were awfully dull; it was his sayings one remembered; his eyes, his pocket-knife, his smile, his grumpiness and, when

millions of things had utterly vanished—how strange it was!—a few sayings like this about cabbages.

8. The passage is reflective of which of the following types of writing?
   a. Persuasive
   b. Expository
   c. Technical
   d. Descriptive
   e. Narrative

9. What was the narrator feeling right before Peter Walsh's voice distracted her?
   a. A spark of excitement for the morning
   b. Anger at the larks
   c. A sense of foreboding
   d. Confusion at the weather
   e. Anger towards men

Sentence Equivalence

Select the two answer choices that can complete the sentence and create sentences that have complementary meaning.

10. The religious group's ________ paved the way for an adversary with just as much extremism and intolerance.
   a. Acumen
   b. Fanaticism
   c. Credulity
   d. Zealotry
   e. Decorum
   f. Sanctity

11. The student's reluctance to speak in front of the classroom due to nervousness was a testimony to her ________.
   a. Reticence
   b. Serendipity
   c. Shrewdness
   d. Virulence
   e. Vocation
   f. Demureness

12. They kept a ________ watch on the family dog after he experienced his first seizure.
   a. Venerable
   b. Superfluous
   c. Temperamental
   d. Vigilant
   e. Fastidious
   f. Recessive

13. There was much needed ________ in the neighborhood after the chaos the hurricane caused over the last few days.

a. Abhorrence
b. Assiduousness
c. Equanimity
d. Irreverence
e. Torpor
f. Aplomb

14. The knowledge of photonic molecules—synthetic matter made by binding photons together to form "molecules"—was ________ knowledge until the twenty-first century.

a. Tenuous
b. Eminent
c. Wayward
d. Somber
e. Esoteric
f. Illicit

15. After his graduation, they celebrated his freedom with a ________ party, a trip to Europe, and a hike of the Appalachian Trail.

a. Morose
b. Convivial
c. Prohibitive
d. Jovial
e. Supercilious
f. Derivative

16. Studies have found that those who smile more often receive more money than those who do not, and also have ________ relationships compared to people who frown more often.

a. Arcane
b. Capricious
c. Viscous
d. Blighted
e. Prolific
f. Auspicious

17. With Christmas just around the corner, shoppers are ________ to get their decorations up and their presents bought and placed under the tree.

a. Ardent
b. Wry
c. Keen
d. Stupefied
e. Adverse
f. Dispelled

Select the best word from the corresponding column of choices that most clearly completes the passage:

18. According to the FDA, those who develop Keshan disease, hypothyroidism, and extreme fatigue, may be experiencing a(n) (i) ________________ of selenium.

Blank (i)
a. Hapless
b. Paucity
c. Incumbent
d. Maelstrom
e. Tout

19. The water pollution (i) __________ the whole city of Harrisburg, causing (ii) ____________ and devastation.

Blank (i)	Blank (ii)
a. Disparaged	d. Pathosis
b. Quelled	e. Oversight
c. Vitiated	f. Propensity

20. The student vowed to (i) __________ her teachers' ban on the cell phone policy by hiding it behind her laptop.

Blank (i)
a. Contravene
b. Rectify
c. Squander
d. Parch
e. Excavate

21. Let us not underestimate the (i) ____________ plants play in our lives. Plants (ii) ________________ oxygen and, during photosynthesis, absorb carbon dioxide. Plants are also a necessity to the human body, (iii) ____________ the appropriate vitamins and minerals needed to make our bodies stronger.

Blank (i)	Blank (ii)	Blank (iii)
a. Aberration	d. Palliate	g. Bestowing
b. Efficacy	e. Elicit	h. Confiscating
c. Listlessness	f. Generate	i. Reaping

22. The country has adopted ecotourism as a way to (i) ____________ poverty and to provide a viable way of spurring economic growth.

Blank (i)
a. Hasten
b. Dupe
c. Mitigate
d. Deride
e. Coalesce

23. In an attempt to (i) _______________ the pain centralized in her back due to a fall three months prior, Shanna called her primary care only to have the nurse tell her that the doctor had (ii) ____________________ for the day.

Blank (i)	Blank (ii)
a. Annex	d. Renounced
b. Covet	e. Disengaged
c. Assuage	f. Precluded

24. As a helping professional, she was there to offer (i) _______________ to those who were in emotional need of it.

Blank (i)
a. Tirade
b. Prattle
c. Folly
d. Solace
e. Dearth

25. The novel was too (i) ______________________ given its exorbitant focus on the uncle's death, the pity evoked by the child's sickness, and the constant state of grief the mother was in.

Blank (i)
a. Lurid
b. Placid
c. Scrupulous
d. Wry
e. Maudlin

Answer Explanations

Section 1

1. D: Neutrality due to the style of the report. The report is mostly objective; we see very little language that entails any strong emotion whatsoever. The story is told almost as an objective documentation of a sequence of actions—we see the president sitting in his box with his wife, their enjoyment of the show, Booth's walk through the crowd to the box, and Ford's consideration of Booth's movements. There is perhaps a small amount of bias when the author mentions the president's "worthy wife." However, the word choice and style show no signs of excitement, sadness, anger, or apprehension from the author's perspective, so the best answer is Choice *D*.

2. B: Mr. Ford assumed Booth's movement throughout the theater was due to being familiar with the theater. Choice *A* is incorrect; although Booth does eventually make his way to Lincoln's box, Mr. Ford does not make this distinction in this part of the passage. Choice *C* is incorrect; although the passage mentions "companions," it mentions Lincoln's companions rather than Booth's companions. Choice *D* is incorrect; the passage mentions "dress circle," which means the first level of the theater, but this is different from a "dressing room." Finally, Choice *E* is incorrect; the passage mentions a "symptom" but does not signify a symptom from an illness.

3. C: A lead singer leaves their band to begin a solo career, and the band drops in sales by 50 percent on their next album. The original source of the analogy displays someone significant to an event who leaves, and then the event becomes worse for it. We see Mr. Sothern leaving the theater company, and then the play becoming a "very dull affair." Choice *A* depicts a dancer who backs out of an event before the final performance, so this is incorrect. Choice *B* shows a basketball player leaving an event, and then the team makes it to the championship but then loses. This choice could be a contestant for the right answer; however, we don't know if the team has become the worst for his departure or the better for it. We simply do not have enough information here. Choice *D* is incorrect. The actor departs an event, but there is no assessment of the quality of the movie. It simply states what actors filled in instead. Choice *E* is incorrect because the opposite of the source happens; the professor leaves the entity, and the entity becomes better. Additionally, the betterment of the entity is not due to the individual leaving. Choice *E* is not analogous to the source.

4. A: A chronological account in a fiction novel of a woman and a man meeting for the first time. It's tempting to mark Choice *A* as incorrect because the genres are different. Choice *A* is a fiction text, and the original passage is not a fictional account. However, the question stem asks specifically for organizational structure. Choice *A* is a chronological structure just like the passage, so this is the correct answer. The passage does not have a cause and effect, problem/solution, or compare/contrast structure, making Choices *B*, *D*, and *E* incorrect. Choice *C* is tempting because it mentions an autobiography; however, the structure of this text starts at the end and works its way toward the beginning, which is the opposite structure of the original passage.

5. B: This is the best answer choice because the author is trying to demonstrate by the examples that anyone who incites a crime, despite the severity or magnitude of the crime, should be held accountable for that crime in some degree. Choice *A* is incorrect because the crimes mentioned are not being compared to each other, but they are being used to demonstrate a point. Choice *C* is incorrect because the author makes the same point using both of the examples and does not question the definition of freedom of speech but its ability to be limited. Choice *D* is incorrect because this sentiment goes against what the author has been arguing throughout the passage. Choice *E* is incorrect because the author does not suggest that the crimes mentioned be reopened anywhere in the passage.

6. A: The idea that human beings are able and likely to change their mind between the utterance and execution of an event that may harm others most seriously undermines the claim because it brings into question the tendency to commit a crime and points out the difference between utterance and action in moral situations. Choice *B* is incorrect; this idea does not undermine the claim at hand, but introduces an observation irrelevant to the claim. Choices *C, D,* and *E* are incorrect because they would actually strengthen the author's claim rather than undermine it.

7. E: The word that would least change the meaning of the sentence is *E*, grievance. Malcontent is a complaint or grievance, and in this context would be uttered in advocation of absolute freedom of speech. Choice *A*, *regimen*, means a pattern of living, and would not make sense in this context. Choice *B*, *cacophony*, means a harsh noise; someone may express or "urge" a cacophony but it would be an awkward word in this context. Choice *C*, *anecdote*, is a short account of an amusing story. Since the word is a noun it fits grammatically inside the sentence, but anecdotes are usually thought out, and this word is considered "unthinking." Choice *D*, *residual*, means to be something of an outcome, or what is left behind or remaining, which does not make sense within this context.

8. B: The author says that the classical understanding of poetry dealt with its ability to be used to teach morality. Later, philosophers would define poetry by its ability to imitate life. Finally, during the renaissance, poetry was believed to be an imitative art that instilled morality in its readers. The rest of the answer choices are mixed together from this explanation in the passage. Poetry was never mentioned for use in entertainment, which makes Choices *D* and *E* incorrect. Choices *A* and *C* are incorrect for mixing up the chronological order.

9. C: That although most poetry was written as lyric, epic, or drama, the critics were most focused on the techniques of the epic and drama and their performance of structure and character. This is the best answer choice as portrayed by paragraph three. Choice *A* is incorrect because nowhere in the passage does it say rhetoric was more valued than poetry, although it did seem to have a more definitive purpose than poetry. Choice *B* is incorrect; this almost mirrors Choice *A,* but the critics were *not* focused on the lyric, as the passage indicates. Choice *D* is incorrect because the passage does not mention that the study of poetics was more pleasurable than the study of rhetoric. Choice *E* is incorrect because, again, we do not see anywhere in the passage where poetry was reserved for the most elite in society.

10. C, D: *Amicable* and *genial* are the correct answer choices because they both denote friendliness. *Belligerent* is incorrect because it means argumentative. *Callous* means cold and unfriendly, so this is the opposite of the correct answer. *Clandestine* means secret or sly, so this is incorrect. Finally, *destitute* is incorrect because it means poor or wanting.

11. B, E: *Enigmatic* and *reticent* are the best choices for this sentence, because they both mean secretive, quiet, or mysterious. *Demonstrative* means expressive, so this is the opposite of what we are looking for. *Apathetic* is incorrect because it means disinterested or uncaring. *Buoyant* means resilient or light in weight and does not fit within the context of the sentence. Finally, *convivial* means fun-loving, so this choice is incorrect.

12. A, B: *Evoke* and *galvanize* are the best answer choices here because they both mean to awaken or arouse something. *Placate* is incorrect because it means to soothe or pacify. *Quell* means to suppress, so this is the opposite of the correct answer. *Raze* is incorrect because it means to flatten or knock down. Finally, *relinquish* is incorrect because it denotes giving up or letting go.

13. C, F: *Reverence* and *deference* are the correct answer choices. *Reverence* means to have a high opinion of something, so this fits within the context. *Deference* means paying homage or attention, so this also fits within the sentence. *Flippancy* is incorrect because it means irreverence, which is the opposite of the correct answer. *Impudence* means audacity, so this is also incorrect. *Indolence* means sloth, which doesn't make sense in this context. Finally, *tenacity* means stubbornness, so this is not the best answer.

14. D, E: *Decimate* and *eradicate* are the correct answer choices because they both mean to destroy or get rid of. *Accost* is incorrect because it means to approach for solicitation or to annoy. *Elicit* is also incorrect because it means to draw out. *Reproach* is not the correct answer because it denotes strong criticism or dishonor, so this does not make sense given the context. *Annex* does not work either because it means an extension or something added.

15. A, C: *Propinquity and rapport* are the best choices here. *Propinquity* means closeness and *rapport* means an understanding between people, so these fit the best. *Rancor* is incorrect because it means bitterness. *Fortitude* means boldness so this is not the best answer choice. *Impasse* is incorrect because it denotes a stalemate or deadlock. *Hubris* is also incorrect because it means arrogance.

16. B, F: *Supercilious* and *patronizing* are correct. *Supercilious* means arrogant or stuck up, while *patronizing* means arrogant or condescending. *Hackneyed* is not the best answer choice because it means clichéd. *Innate* means native or natural, so this is incorrect. *Prudent* means careful, so this is also incorrect. *Subtle* is not the correct answer choice because it means subdued or muted.

17. D, F: *Opulent* and *palatial* are the correct choices because they both denote luxury. *Frugal* means economical so this is the opposite of the correct answer. *Hardy* means tough so this is incorrect. *Inane* is incorrect because it means stupid. Finally, *lucrative* means well-paid, not expensive, so this is not the best answer choice.

18. B, D, I: For (i), Choice *B*, *affluent*, is the correct answer because it means wealth. *Destitute* means poor, and *hapless* means unfortunate or deserving pity, so these are incorrect. For (ii), Choice *D*, *migrate*, is the correct answer because it means to travel. *Exhort* means to encourage by cheers or shouts, and *rescind* means to officially cancel something, so these terms are incorrect. For (iii), Choice *I*, *lucrative*, is the best answer because it means productive or well-paid. *Puerile* means a lack of maturity and *garrulous* means full of trivial conversation, so these are incorrect.

19. A, F: For (i), Choice *A*, *equestrian*, is the correct answer because it means having to do with horses. *Equanimity* means levelheadedness, and *flagrant* means without shame or reprehensible, so these are incorrect. For (ii), Choice *F*, *consternation*, means dismay, so we can mark this as correct. *Blandishment* means flattery and *mendaciousness* means given to lying, so these are incorrect.

20. C: The best answer is *proclivity*, Choice *C*, because it means having a natural inclination towards something. *Tirade* is incorrect because it means a violent speech. *Requisition* is incorrect because it means demand. *Negligence* is incorrect because it means lack of attention. Finally, *irreverence* is treating something with disrespect, so this is also incorrect.

21. A, E: For (i), the best answer choice is *A*, *impetuous*, because it means something done hastily and with a lack of thought. *Ambivalent* is incorrect because it means unable to decide on something. *Fortuitous* is incorrect because it means happy or lucky. For (ii), *concatenation*, Choice *E*, is the correct answer because it means sequence. *Fetter* is incorrect because it means shackles. *Turpitude* is incorrect because it means a corrupt act and does not make sense in this context.

22. B: The correct answer is Choice *B*, *adept*, because it means skilled. *Banal* is incorrect because it means common and does not quite fit within the context. *Divisive* is incorrect because it means to create conflict. *Indignant* also doesn't make sense here because it means angry. Finally, *precarious* is incorrect because it means dangerous.

23. C, D: For (i), Choice *C* is the correct answer, because *rescinded* means to take away or retract. Choice *A*, sage, means wise, so this is incorrect. Choice *B*, *benevolent*, means generous, so this is incorrect. For (ii), *punctilious* is the correct answer because it means meticulous. The word "otherwise" should indicate that contrast from her previous mistake. *Bombastic* means pompous and *disputatious* means quarrelsome, so these are considered incorrect.

24. E: *Didactic* is the correct answer because it means to provide a moral or lesson of some sort. *Elusive* means difficult to catch or define and is not the correct answer choice. *Indulgent* means being lenient or considerate, and this also does not make sense in the context of the sentence. *Morose* is incorrect because it means depressive, and nothing in the context lets us know the play was depressive. Finally, *tenuous* is incorrect because it means insignificant, and we are looking for something that makes people uncomfortable and teaches "lessons."

25. A, E: For (i), *archetypal* is the best answer because it means a typical example of something. *Amenable* means yielding and *antediluvian* means very old, so these answers are incorrect. For (ii), *circumvent* is the correct answer because it means to avoid or escape. *Coalesce* means to fuse together or blend, so this doesn't make sense in the context of the sentence. *Encompass* is also incorrect because it means to surround.

Section 2

1. D: The word *deleterious* can be best interpreted as referring to the word *ruinous*. The first paragraph attempts to explain the process of milk souring, so the "acid" would probably prove "ruinous" to the growth of bacteria and cause souring. Choice *A*, *amicable*, means friendly, so this does not make sense in context. Choice *B*, *smoldering*, means to boil or simmer, so this is also incorrect. Choices *C* and *E*, *luminous* and *virtuous*, have positive connotations and don't make sense in the context of the passage. *Luminous* means shining or brilliant, and *virtuous* means to be honest or ethical.

2. B: The author begins by explaining a process or phenomenon, then gives the history of the study of this phenomenon, then ends by presenting the effects of this phenomenon. The author explains the process of souring in the first paragraph by informing the reader that "it is due to the action of certain of the milk bacteria upon the milk sugar which converts it into lactic acid, and this acid gives the sour taste and curdles the milk." In paragraph two, we see how the phenomenon of milk souring was viewed when it was "first studied," and then we proceed to gain insight into "recent investigations" toward the end of the paragraph. Finally, the passage ends by presenting the effects of the phenomenon of milk souring. We see the milk curdling, becoming bitter, tasting soapy, turning blue, or becoming thread-like. All of the other answer choices are incorrect.

3: A: To inform the reader of the phenomenon, investigation, and consequences of milk souring. Choice *B* is incorrect because the passage states that *Bacillus acidi lactici* is not the only cause of milk souring. Choice *C* is incorrect because, although the author mentions the findings of researchers, the main purpose of the text does not seek to describe their accounts and findings, as we are not even told the names of any of the researchers. Choice *D* is tricky. We do see the author present us with new findings in contrast to the first cases studied by researchers. However, this information is only in the second paragraph, so it is not the primary purpose of the entire passage. Finally, Choice *E* is incorrect because the genre of the passage is more informative than narrative, although the author does talk about the phenomenon of milk souring and its subsequent effects.

4. C: Milk souring is caused mostly by a species of bacteria identical to that of *Bacillus acidi lactici* although there are a variety of other bacteria that cause milk souring as well. Choice *A* is incorrect because it contradicts the assertion that the souring is still caused by a variety of bacteria. Choice *B* is incorrect because the ordinary cause of milk souring *is known* to current researchers. Choice *D* is incorrect because this names mostly the effects of milk souring, not the cause. Choice *E* is incorrect because the bacteria itself doesn't have a strange soapy smell or is a different color, but it eventually will cause the milk to produce these effects.

5. D: To call upon the interpretation of freedom of speech to be already evident in the First Amendment and to offer a clear perimeter of the principle during war time. Choice *A* is incorrect; the passage calls upon no historical situations as precedent in this passage. Choice *B* is incorrect; we can infer that the author would not agree with this, because they state that "In war time, therefore, speech should be unrestricted . . . by punishment." Choice *C* is incorrect; this is more consistent with the main idea of the first passage. Choice *E* is incorrect; the passage states a limitation in saying that "speech should be unrestricted . . . unless it is clearly liable to cause direct and dangerous interference with the conduct of war."

6. A: To show the audience one of the effects of criminal rehabilitation by comparison. Choice *B* is incorrect because although it is obvious the author favors rehabilitation, the author never asks for donations from the audience. Choices *C* and *D* are also incorrect. We can infer from the passage that American prisons are probably harsher than Norwegian prisons. However, the best answer that captures the author's purpose is Choice *A*, because we see an effect by the author (recidivism rate of each country) comparing Norwegian and American prisons. Choice *E* is also incorrect. Although we see a description of rehabilitative prisons at the end of the paragraph, this does not capture the entire purpose of the passage.

7. D: The likelihood of a convicted criminal to reoffend. The passage explains how a Norwegian prison, due to rehabilitation, has a smaller rate of recidivism. Thus, we can infer that recidivism is probably not a positive attribute. Choices *A* and *B* are both positive attributes, the lack of violence and the opportunity of inmates to receive therapy, so Norway would probably not have a lower rate of these two things. Choice *C* is possible, but it does not make sense in context, because the author does not talk about tactics to keep prisoners inside the compound, but ways in which to rehabilitate criminals so that they can live as citizens when they get out of prison. Choice *E* would be best described as *refutation*, not *recidivism.*

8. E: The passage is reflective of a narrative. A narrative is used to tell a story, as we see the narrator trying to do in this passage by using memory and dialogue. Choice *A*, persuasive writing, uses rhetorical devices to try to convince the audience of something, and there is no persuasion or argument within this passage. Choice *B*, expository, is a type of writing used to inform the reader. Choice *C*, technical writing, is usually used within business communications and uses technical language to explain procedures or concepts to someone within the same technical field. Choice *D*, descriptive writing, is a type of writing that paints a picture of people or setting using concrete words that play on the audience's five senses.

9. C: A sense of foreboding. The narrator, after feeling excitement for the morning, feels "that something awful was about to happen," which is considered foreboding. The narrator mentions larks and weather in the passage, but there is no proof of anger or confusion at either of them. The narrator does not express anger towards men before being distracted.

10. B, D: The correct answers are *fanaticism* and *zealotry*, which both denote extremism or excessive zeal. Choice *A*, *acumen*, means insightfulness, so this can be marked as incorrect. Choice *C*, *credulity*, means trusting others too much, so this also does not make sense. Choice *E*, *decorum*, is also incorrect, as it means order and politeness. Finally, Choice *F*, *sanctity*, though a religious word, does not fit logically into the context of the sentence. This is incorrect.

11. A, F: These are the best answers because they both mean shyness or reserve. Choice *B*, *serendipity*, means *luck*, and does not fit within the context of the sentence. Choice *C*, shrewdness, also does not make sense here, as shrewdness means smart or astute. Choice *D*, *virulence*, means resentment, so this is incorrect. Finally Choice *E*, *vocation*, is incorrect because it means career.

12. D, E: *Vigilant* means watchful and alert while *fastidious* means meticulous, so these words are able to be interchangeable in the context of the sentence. Choice *A*, venerable, is incorrect because it means deserving respect due to age. Choice *B*, *superfluous*, means more than necessary, so this is incorrect. Choice *C*, *temperamental*, means moody, so we can mark this off as incorrect. Finally, Choice *F*, *recessive*, is incorrect because it means passive.

13. C, F: The best answer choices are *equanimity* and *aplomb*, because they both denote levelheadedness and calmness. We are looking for something that contrasts "chaos." *Abhorrence* is incorrect because it means hatred. *Assiduousness* isn't the opposite of calm; it means diligence. However, this is not the *best* answer, so mark it as incorrect. *Irreverence* means disrespect so this is incorrect. Finally, *torpor* means drowsiness, so this is incorrect.

14. A, E: The correct answers are *tenuous* and *esoteric*, which both have something to do with having weak or mysterious information that is not totally whole or complete. *Eminent* is incorrect because it means prominent or renowned, which is the opposite of what we are looking for. *Wayward* means stubborn or selfish, so this is also incorrect. *Somber* is incorrect because it means bleak. Finally, *illicit* is incorrect because it means illegal, and this doesn't make sense in the context of the sentence.

15. B, D: *Convivial* and *jovial* are the correct answers because they both mean fun-loving and friendly. *Morose* is incorrect because it means cranky or melancholy, and there's no hint that the party went sour. *Prohibitive* is incorrect because it means restrictive. *Supercilious* is also incorrect because it means arrogant. Finally, *derivative* means copied or borrowed, so this is incorrect.

16. E, F: *Prolific* and *auspicious* are the best choices for this sentence. *Prolific* means fruitful and productive, while *auspicious* means encouraging and favorable, both which match the tone of the sentence. *Arcane* means secret or hidden, so this is incorrect. *Capricious* means arbitrary or careless, so this would not make sense given the context. *Viscous* means sticky or gummy, so this is especially irrelevant to the sentence. Finally, *blighted* means ruin or destroy, so this is also incorrect.

17. A, C: *Ardent* and *keen* are the best words to use for this sentence because they both mean to be enthusiastic about something. *Wry* means clever or grim humor, so this choice does not make sense in this context. *Stupefied* means bewildered, so this is incorrect. *Adverse* is not the best option here because it means opposing, so mark this as incorrect. Finally, *dispelled* is incorrect because it means to go in different directions.

18. B: *Paucity* is the correct answer because it means lack of something or insufficiency. There is a lack of selenium, in this case. *Hapless* is incorrect because it means deserving pity, which doesn't make sense in this context. *Incumbent* is incorrect because it means morally binding. *Maelstrom* means agitation or chaos, so this is incorrect. Finally, *tout* is incorrect because it means praise, which doesn't make sense in this context.

19. C, D: For (i), *vitiated* is the correct answer because it means to spoil or impair the quality of something. *Disparaged* is incorrect because it means to belittle, so this doesn't make sense. *Quelled* is also incorrect because it means to quench or suppress something. For (ii), *pathosis* is the correct answer because it means a diseased condition. *Oversight* means mistake, so this is not the best answer. Finally, *propensity* is incorrect because it means inclination or tendency.

20. A: *Contravene* is the correct answer because it means to defy or go against a rule. *Rectify* is incorrect because it means to make something right. *Squander* is incorrect, because it means to waste, so this does not make sense in the context of the sentence. *Parch* means to make dry, so mark this as incorrect. Finally, *excavate* means to dig up, so this is also incorrect.

21. B, F, G: For (i), *efficacy* is the correct answer because it means efficiency and productiveness. *Aberration* means deviation, so this choice is incorrect. *Listlessness* is incorrect because it means sluggish. For (ii), *generate* is the correct answer because it means to produce or create. *Elicit* means to call forth or draw out—this is close, but usually is not the word used for plants, so mark it as incorrect. *Palliate* is incorrect because it means to cover up or gloss over. For (iii), *bestowing* is the correct answer because it means giving or allotting. *Confiscating* is incorrect because it means to steal or seize, which is the opposite of the correct answer. Finally, *reaping* is incorrect because it means to gather or harvest.

22. C: *Mitigate* is the correct answer because it means to alleviate something, in this case, poverty. *Hasten* is incorrect because it means to speed something up. *Dupe* means deceive, so this answer choice is also incorrect given the context. *Deride* is also incorrect because it means to make fun of something. *Coalesce* means to blend or fuse, so this is also incorrect.

23. C, E: For (i), *assuage* means to relieve or appease, in this case Shanna's pain, so Choice *C* is the best choice. *Annex* is incorrect because it means attach to. *Covet* is also incorrect because it means to long for something someone else has. For (ii), *disengaged* is the correct answer because it means to detach or withdraw. *Renounced*, which means reject, doesn't make sense in this context, so mark this as incorrect. *Precluded* is incorrect because it means to make impossible.

24. D: *Solace* is the correct answer. Helping professionals provide solace to those who are in need of it. Solace means comfort or peace. *Tirade* is incorrect because it is speech that is intended to be critical. *Prattle* means babble, and although prattle is sometimes used as a synonym for talk, and helping professionals do talk to clients, this is not the best answer given the word choice and the context. *Folly* means foolishness, so this is incorrect. *Dearth* is incorrect because it means scarcity.

25. E: Choice *E*, *maudlin*, is the correct answer, because it means oversentimental. Choice *A* is incorrect; *lurid* means sensational or shocking, so this isn't the best answer choice given the examples of sentiment and grief. Choice *B*, placid, means calm, so this is incorrect. *Scrupulous* means precise, so Choice *C* is also incorrect. Finally, *wry* means to have a grim sense of humor, so this is also incorrect.

Quantitative Reasoning

Arithmetic

Properties and Types of Integers

Whole numbers are the numbers 0, 1, 2, 3, ... Examples of other whole numbers would be 413 and 8,431. Notice that numbers such as 4.13 and $\frac{1}{4}$ are not included in whole numbers. **Counting numbers**, also known as **natural numbers**, consist of all whole numbers except for the zero. In set notation, the natural numbers are the set $\{1, 2, 3, \dots\}$. The entire set of whole numbers and negative versions of those same numbers comprise the set of numbers known as integers. Therefore, in set notation, the integers are $\{\dots, -3, -2, -1, 0, 1, 2,3, \dots\}$. Examples of other integers are −4,981 and 90,131.

Fractions are parts of a whole. For example, if an entire pie is cut into two pieces, each piece is half of the pie, or $\frac{1}{2}$. The top number in any fraction, known as the **numerator,** defines how many parts there are. The bottom number, known as the **denominator,** states how many pieces the whole is divided into. Fractions can also be negative or written in their corresponding decimal form.

A **decimal** is a number that uses a decimal point and numbers to the right of the decimal point representing the part of the number that is less than 1. For example, 3.5 is a decimal and is equivalent to the fraction $\frac{7}{2}$ or the mixed number $3\frac{1}{2}$. The decimal is found by dividing 2 into 7. Other examples of fractions are $\frac{2}{7}, \frac{-3}{14}$, and $\frac{14}{27}$.

Any number that can be expressed as a fraction is known as a **rational number.** Basically, if a and b are any integers and $b \neq 0$, then $\frac{a}{b}$ is a rational number. Any integer can be written as a fraction where the denominator is 1, so therefore rational numbers consist of all fractions and all integers.

Any number that is not rational is known as an **irrational number.** Consider the number

$$\pi = 3.141592654 \dots.$$

The decimal portion of that number extends indefinitely. In that situation, a number can never be written as a fraction. Another example of an irrational number is $\sqrt{2} = 1.414213662 \dots$. Again, this number cannot be written as a ratio of two integers.

Together, the set of all rational and irrational numbers makes up the real numbers.

Divisibility

A number is **divisible** by another number if when the division is performed, the result is a whole number. For example, 48 is divisible by 8 because $48 \div 8 = 6$, a whole number. A number is divisible by 2 if it is even. A number is divisible by 5 if it ends in a 0 or a 5. Finally, a number is divisible by 3 if the sum of all its digits is divisible by 3.

Factorization

Factorization is the process of breaking up a mathematical quantity, such as a number or polynomial, into a product of two or more factors. For example, a factorization of the number 16 is $16 = 8 \times 2$. If multiplied out, the factorization results in the original number. A **prime factorization** is a specific factorization when the number is factored completely using prime numbers only. For example, the prime factorization of 16 is:

$$16 = 2 \times 2 \times 2 \times 2$$

A factor tree can be used to find the prime factorization of any number. Within a factor tree, pairs of factors are found until no other factors can be used, as in the following factor tree of the number 84:

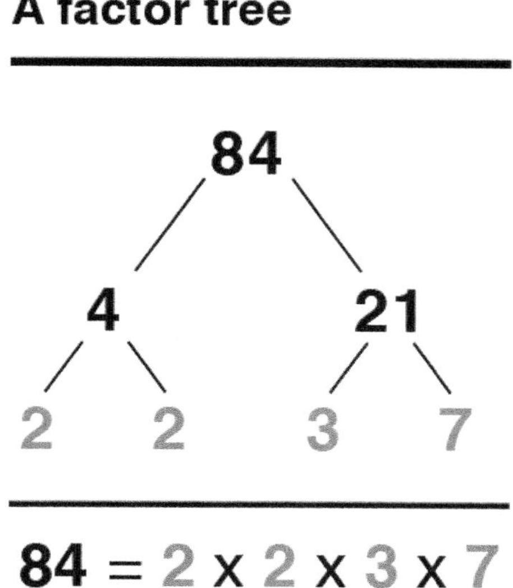

It first breaks 84 into 21×4, which is not a prime factorization. Then, both 4 and 21 are factored into their primes. The final numbers on each branch consist of the numbers within the prime factorization. Therefore,

$$84 = 2 \times 2 \times 3 \times 7$$

Factorization can be helpful in finding greatest common divisors and least common denominators.

Prime Numbers

Prime numbers consist of natural numbers greater than 1 that are not divisible by any other natural numbers other than themselves and 1. For example, 3, 5, and 7 are prime numbers. If a natural number is not prime, it is known as a composite number. 8 is a composite number because it is divisible by both 2 and 4, which are natural numbers other than itself and 1.

Odd and Even Integers

Even numbers are integers that are divisible by 2. For example, 6, 100, 0, and −200 are all even numbers. **Odd** numbers are integers that are not divisible by 2. If an odd number is divided by 2, the result is a fraction. For example, −5, 11, and −121 are odd numbers.

Arithmetic Operations

The four basic operations include addition, subtraction, multiplication, and division. The result of **addition** is a sum, the result of **subtraction** is a difference, the result of **multiplication** is a product, and the result of **division** is a quotient. Each type of operation can be used when working with rational numbers; however, the basic operations need to be understood first while using simpler numbers before working with fractions and decimals.

These operations should first be learned using whole numbers. Addition needs to be done column by column. To add two whole numbers, add the ones column first, then the tens columns, then the hundreds, etc. If the sum of any column is greater than 9, a one must be carried over to the next column. For example, the following is the result of 482+924:

$$\begin{array}{r} {\scriptstyle 1} \\ 482 \\ +924 \\ \hline 1406 \end{array}$$

Notice that the sum of the tens column was 10, so a one was carried over to the hundreds column. Subtraction is also performed column by column. Subtraction is performed in the ones column first, then the tens, etc. If the number on top is less than the number below, a one must be borrowed from the column to the left. For example, the following is the result of 5,424 – 756:

$$\begin{array}{rrrr} 4 & 13 & 11 & 14 \\ \not{5} & \not{4} & \not{2} & \not{4} \\ - & 7 & 5 & 6 \\ \hline 4 & 6 & 6 & 8 \end{array}$$

Notice that a one is borrowed from the tens, hundreds, and thousands place. After subtraction, the answer can be checked through addition. A check of this problem would be to show that:

$$756 + 4{,}668 = 5{,}424$$

In multiplication, the number on top is known as the **multiplicand**, and the number below is the **multiplier**. Complete the problem by multiplying the multiplicand by each digit of the multiplier. Make sure to place the ones value of each result under the multiplying digit in the multiplier. Each value to the right is then a 0. The product is found by adding each product. The following example shows the process of multiplying 46 times 37:

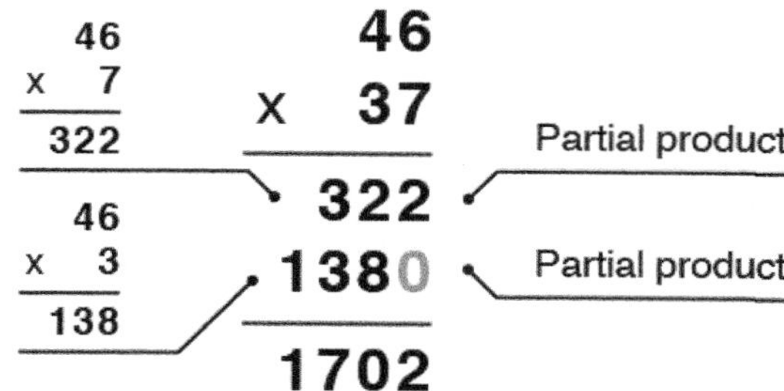

Finally, division can be performed using long division. When dividing, the first number is known as the **dividend,** and the second is the **divisor**. For example, with $a \div b = c$, a is the dividend, b is the divisor, and c is the quotient. For long division, place the dividend within the division bar and the divisor on the outside. For example, with $8{,}764 \div 4$, refer to the first problem in the diagram on the next page. First, there are two 4's in the first digit, 8. This number 2 gets written above the 8. Then, multiply 4 times 2 to get 8, and that product goes below the 8. Subtract to get 0, and then carry down the second digit, 7.

Continue the same steps. $7 \div 4 = 1$ R3, so 1 is written above the 7.

Multiply 4 times 1 to get 4, and write it below the 7.

Subtract to get 3, and carry the 6 down next to the 3.

Continuing this process for the next two digits results in a 9 and a 1. The final subtraction results in a 0, which means that 8,764 is evenly divisible by 4 with no remaining numbers.

The second example shows that

$$4{,}536 \div 216 = 21$$

The steps are a little different because 216 cannot be contained in 4 or 5, so the first step is placing a 2 above the 3 because there are two 216's in 453. Finally, the third example shows that

$$546 \div 31 = 17 \text{ R}19$$

The 19 is a **remainder**. Notice that the final subtraction does not result in a 0, which means that 546 is not divisible by 31. The remainder can also be written as a fraction over the divisor to say that

$$546 \div 31 = 17\frac{19}{31}$$

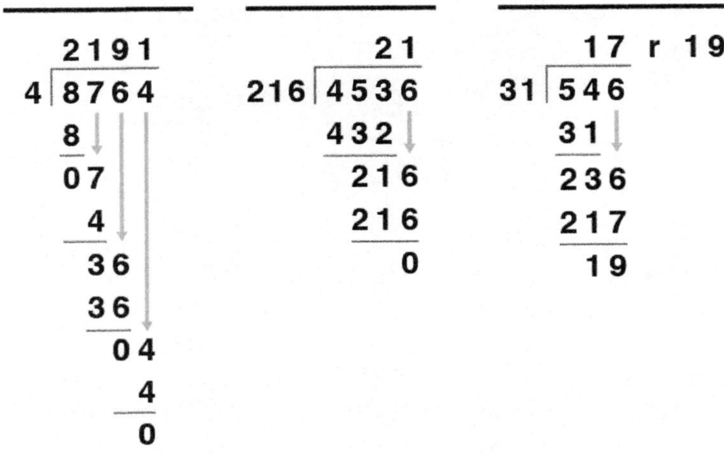

A remainder can have meaning in a division problem with real-world application. For example, consider the third example,

$$546 \div 31 = 17 \text{ R}19$$

Let's say that we had $546 to spend on calculators that cost $31 each, and we wanted to know how many we could buy. The division problem would answer this question. The result states that 17 calculators could be purchased, with $19 left over. Notice that the remainder will never be greater than or equal to the divisor.

Once the operations are understood with whole numbers, they can be used with negative numbers. There are many rules surrounding operations with negative numbers. First, consider addition with integers. The sum of two numbers can first be shown using a number line.

For example, to add $-5 + (-6)$, plot the point −5 on the number line. Adding a negative number is the same as subtracting, so move 6 units to the left. This process results in landing on −11 on the number line, which is the sum of −5 and −6. If adding a positive number, move to the right. While visualizing this process using a number line is useful for understanding, it is more efficient to learn the rules of operations.

When adding two numbers with the same sign, add the absolute values of both numbers, and use the common sign of both numbers as the sign of the sum.

For example, to add $-5 + (-6)$, add their absolute values $5 + 6 = 11$.

Then, introduce a negative number because both addends are negative. The result is −11.

To add two integers with unlike signs, subtract the lesser absolute value from the greater absolute value, and apply the sign of the number with the greater absolute value to the result.

For example, the sum $-7 + 4$ can be computed by finding the difference $7 - 4 = 3$ and then applying a negative because the value with the larger absolute value is negative. The result is −3.

Similarly, the sum $-4 + 7$ can be found by computing the same difference but leaving it as a positive result because the addend with the larger absolute value is positive. Also, recall that any number plus 0 equals that number.

This is known as the **Addition Property of 0.**

Subtracting two integers with opposite signs can be computed by changing to addition to avoid confusion. The rule is to add the first number to the opposite of the second number. The opposite of a number is the number with the same value on the other side of 0 on the number line.

For example, −2 and 2 are opposites. Consider $4 - 8$. Change this to adding the opposite as follows: $4 + (-8)$.

Then, follow the rules of addition of integers to obtain −4.

Secondly, consider $-8 - (-2)$. Change this problem to adding the opposite as $-8 + 2$, which equals −6. Notice that subtracting a negative number functions the same as adding a positive number.

Multiplication and division of integers are actually less confusing than addition and subtraction because the rules are simpler to understand.

If two factors in a multiplication problem have the same sign, the result is positive. If one factor is positive and one factor is negative, the result, known as the **product,** is negative.

For example,

$$(-9)(-3) = 27$$

and

$$9(-3) = -27$$

Also, any number times 0 always results in 0. If a problem consists of several multipliers, the result is negative if it contains an odd number of negative factors, and the result is positive if it contains an even number of negative factors. For example:

$$(-1)(-1)(-1)(-1) = 1$$

and

$$(-1)(-1)(-1)(-1)(-1) = -1$$

Similar rules apply within division. First, consider some vocabulary. When dividing 14 by 2, it can be written in the following ways:

$$14 \div 2 = 7$$

or

$$\frac{14}{2} = 7$$

14 is the **dividend,** 2 is the **divisor,** and 7 is the **quotient.** If two numbers in a division problem have the same sign, the quotient is positive. If two numbers in a division problem have different signs, the quotient is negative. For example:

$$14 \div (-2) = -7$$

and

$$-14 \div (-2) = 7$$

To check division, multiply the quotient times the divisor to obtain the dividend. Also, remember that 0 divided by any number is equal to 0. However, any number divided by 0 is undefined. It just does not make sense to divide a number by 0 parts.

If more than one operation is to be completed in a problem, follow the **Order of Operations**. The mnemonic device, **PEMDAS**, states the order in which addition, subtraction, multiplication, and division need to be done.

It also includes when to evaluate operations within grouping symbols and when to incorporate exponents. PEMDAS, which some remember by thinking "please excuse my dear Aunt Sally," refers to parentheses, exponents, multiplication, division, addition, and subtraction.

First, complete any operation within parentheses or any other grouping symbol like brackets, braces, or absolute value symbols. Note that this does not refer to when parentheses are used to represent multiplication like $(2)(5)$. An operation is not within parentheses like it is in (2×5). Then, any exponents must be computed. Next, multiplication and division are performed from left to right. Finally, addition and subtraction are performed from left to right.

The following is an example in which the operations within the parentheses need to be performed first, so the order of operations must be applied to the exponent, subtraction, addition, and multiplication within the grouping symbol:

$$9-3(3^2-3+4\cdot 3)$$

$$\begin{aligned} & 9-3(3^2-3+4\cdot 3) \quad \text{Work within the parentheses first} \\ &= 9-3(9-3+12) \\ &= 9-3(18) \\ &= 9-54 \\ &= -45 \end{aligned}$$

Once the rules for integers are understood, move on to learning how to perform operations with fractions and decimals. Recall that a rational number can be written as a fraction and can be converted to a decimal through division.

If a rational number is negative, the rules for adding, subtracting, multiplying, and dividing integers must be used. If a rational number is in fraction form, performing addition, subtraction, multiplication, and division is more complicated than when working with integers. First, consider addition. To add two fractions having the same denominator, add the numerators and then reduce the fraction. When an answer is a fraction, it should always be in lowest terms.

Lowest terms means that every common factor, other than 1, between the numerator and denominator is divided out. For example:

$$\frac{2}{8}+\frac{4}{8}=\frac{6}{8}=\frac{6\div 2}{8\div 2}=\frac{3}{4}$$

Both the numerator and denominator of $\frac{6}{8}$ have a common factor of 2, so 2 is divided out of each number to put the fraction in lowest terms. If denominators are different in an addition problem, the fractions must be converted to have common denominators.

The **least common denominator (LCD)** of all the given denominators must be found, and this value is equal to the **least common multiple (LCM)** of the denominators. This non-zero value is the smallest number that is a multiple of both denominators. Then, rewrite each original fraction as an equivalent fraction using the new denominator. Once in this form, apply the process of adding with like denominators.

For example, consider $\frac{1}{3}+\frac{4}{9}$. The LCD is 9 because it is the smallest multiple of both 3 and 9.

The fraction $\frac{1}{3}$ must be rewritten with 9 as its denominator. Therefore, multiply both the numerator and denominator times 3. Multiplying times $\frac{3}{3}$ is the same as multiplying times 1, which does not change the value of the fraction. Therefore, an equivalent fraction is $\frac{3}{9}$, and

$$\frac{1}{3}+\frac{4}{9}=\frac{3}{9}+\frac{4}{9}=\frac{7}{9}$$

which is in lowest terms. Subtraction is performed in a similar manner; once the denominators are equal, the numerators are then subtracted. The following is an example of addition of a positive and a negative fraction:

$$-\frac{5}{12}+\frac{5}{9}=-\frac{5\times 3}{12\times 3}+\frac{5\times 4}{9\times 4}=-\frac{15}{36}+\frac{20}{36}=\frac{5}{36}$$

Common denominators are not used in multiplication and division. To multiply two fractions, multiply the numerators together and the denominators together. Then, write the result in lowest terms. For example:

$$\frac{2}{3}\times\frac{9}{4}=\frac{18}{12}=\frac{3}{2}$$

Alternatively, the fractions could be factored first to cancel out any common factors before performing the multiplication. For example:

$$\frac{2}{3}\times\frac{9}{4}=\frac{2}{3}\times\frac{3\times 3}{2\times 2}=\frac{3}{2}$$

This second approach is helpful when working with larger numbers, as common factors might not be obvious. Multiplication and division of fractions are related because the division of two fractions is changed into a multiplication problem. This means that dividing a fraction by another fraction is the same as multiplying the first fraction by the reciprocal of the second fraction, so that second fraction must be inverted, or "flipped," to be in reciprocal form. For example:

$$\frac{11}{15}\div\frac{3}{5}=\frac{11}{15}\times\frac{5}{3}=\frac{55}{45}=\frac{11}{9}$$

The fraction $\frac{5}{3}$ is the reciprocal of $\frac{3}{5}$. It is possible to multiply and divide numbers containing a mix of integers and fractions. In this case, convert the integer to a fraction by placing it over a denominator of 1. For example, a division problem involving an integer and a fraction is:

$$3\div\frac{1}{2}=\frac{3}{1}\times\frac{2}{1}=\frac{6}{1}=6$$

Finally, when performing operations with rational numbers that are negative, the same rules apply as when performing operations with integers. For example, a negative fraction times a negative fraction results in a positive value, and a negative fraction subtracted from a negative fraction results in a negative value.

Operations can be performed on rational numbers in decimal form. Recall that to write a fraction as an equivalent decimal expression, divide the numerator by the denominator. For example:

$$\frac{1}{8} = 1 \div 8 = 0.125$$

With the case of decimals, it is important to keep track of place value. To add decimals, make sure the decimal places are in alignment and add vertically. If the numbers do not line up because there are extra or missing place values in one of the numbers, then zeros may be used as placeholders. For example, $0.123 + 0.23$ becomes:

$$\begin{array}{r} 0.123 \\ +\underline{0.230} \\ 0.353 \end{array}$$

Subtraction is done the same way. Multiplication and division are more complicated. To multiply two decimals, place one on top of the other as in a regular multiplication process and do not worry about lining up the decimal points. Then, multiply as with whole numbers, ignoring the decimals. Finally, in the solution, insert the decimal point as many places to the left as there are total decimal values in the original problem. Here is an example of a decimal multiplication problem:

$$\begin{array}{rl} 0.52 & \textit{2 decimal places} \\ \underline{\times\ \ 0.2} & \textit{1 decimal place} \\ 0.104 & \textit{3 decimal places} \end{array}$$

The answer to 52 times 2 is 104, and because there are three decimal values in the problem, the decimal point is positioned three units to the left in the answer.

The decimal point plays an integral role throughout the whole problem when dividing with decimals. First, set up the problem in a long division format. If the divisor is not an integer, move the decimal to the right as many units as needed to make it an integer. The decimal in the dividend must be moved to the right the same number of places to maintain equality. Then, complete division normally.

Here is an example of long division with decimals:

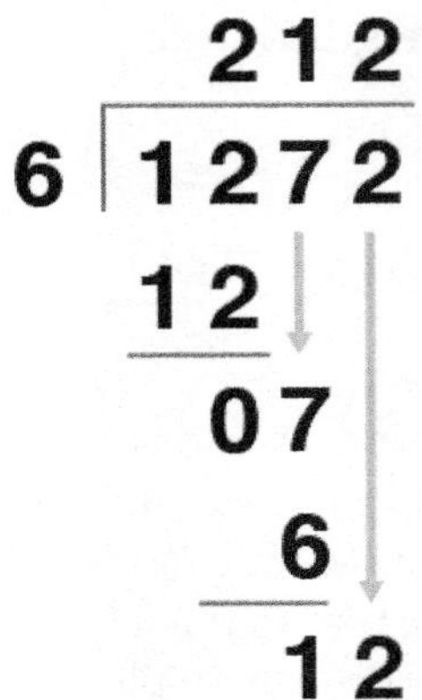

The decimal point in 0.06 needed to move two units to the right to turn it into an integer (6), so it also needed to move two units to the right in 12.72 to make it 1,272. The quotient is 212. To check a division problem, multiply the answer by the divisor to see if the result is equal to the dividend.

Sometimes it is helpful to round answers that are in decimal form. First, find the place to which the rounding needs to be done. Then, look at the digit to the right of it. If that digit is 4 or less, the number in the place value to its left stays the same, and everything to its right becomes a 0. This process is known as **rounding down**. If that digit is 5 or higher, the number in the place value to its left increases by 1, and every number to its right becomes a 0. This is called rounding up. Excess 0s at the end of a decimal can be dropped. For example, 0.145 rounded to the nearest hundredth place would be rounded up to 0.15, and 0.145 rounded to the nearest tenth place would be rounded down to 0.1.

Exponents

Another arithmetic concept is the use of repeated multiplication, which can be written in a more compact notation using **exponents**. The guide previously mentioned the following examples:

$$(-1)(-1)(-1)(-1) = 1$$

and

$$(-1)(-1)(-1)(-1)(-1) = -1$$

The first example can be written as $(-1)^4 = 1$, and the second example can be written as:

$$(-1)^5 = -1$$

Both are exponential expressions; −1 is the base in both instances, and 4 and 5 are the respective exponents. Note that a negative number raised to an odd power is always negative, and a negative number raised to an even power is always positive. Also, $(-1)^4$ is not the same as -1^4. In the first expression, the negative is included in the parentheses, but it is not in the second expression. The second expression is found by evaluating 1^4 first to get 1 and then by applying the negative sign to obtain −1.

Roots

Another operation that can be performed on rational numbers is the **square root**. Dealing with real numbers only, the positive square root of a number is equal to one of the two repeated positive factors of that number. For example:

$$\sqrt{49} = \sqrt{7 \times 7} = 7$$

A **perfect square** is a number that has a whole number as its square root. Examples of perfect squares are 1, 4, 9, 16, 25, etc. If a number is not a perfect square, an approximation can be used with a calculator. For example,

$$\sqrt{67} = 8.185$$

rounded to the nearest thousandth place. Taking the square root of a fraction that includes perfect squares involves breaking up the problem into the square root of the numerator separate from the square root of the denominator. For example:

$$\sqrt{\frac{16}{25}} = \frac{\sqrt{16}}{\sqrt{25}} = \frac{4}{5}$$

If the fraction does not contain perfect squares, a calculator can be used. Therefore, $\sqrt{\frac{2}{5}} = 0.632$, rounded to the nearest thousandth place.

In addition to the square root, the **cube root** is another operation. If a number is a **perfect cube,** the cube root of that number is equal to one of the three repeated factors. For example:

$$\sqrt[3]{27} = \sqrt[3]{3 \times 3 \times 3} = 3$$

A negative number has a cube root, which will also be a negative number. For example:

$$\sqrt[3]{-27} = \sqrt[3]{(-3)(-3)(-3)} = -3$$

Similar to square roots, if the number is not a perfect cube, a calculator can be used to find an approximation. Therefore, $\sqrt[3]{\frac{2}{3}} = 0.873$, rounded to the nearest thousandth place.

Higher-order roots also exist. The number relating to the root is known as the **index**. Given the following root, $\sqrt[3]{64}$, 3 is the index, and 64 is the **radicand**. The entire expression is known as the **radical. Higher-order roots** exist when the index is larger than 3. They can be broken up into two groups: even and odd roots. **Even roots**, when the index is an even number, follow the properties of square roots. They are found by finding the number that, when multiplied by itself the number of times indicated by the index, results in the radicand. For example, the fifth root of 32 is equal to 2 because:

$$\sqrt[5]{32} = \sqrt[5]{2 \times 2 \times 2 \times 2 \times 2} = 2$$

Odd roots, when the index is an odd number, follow the properties of cube roots. A negative number has an odd root. Similarly, an odd root is found by finding the single factor that is repeated that many times to obtain the radicand. For example, the 4th root of 81 is equal to 3 because $3^4 = 81$. This radical is written as:

$$\sqrt[4]{81} = 3$$

Higher-order roots can also be evaluated on fractions and decimals, for example, because $\left(\frac{2}{7}\right)^4 = \frac{16}{2,401}$, $\sqrt[4]{\frac{16}{2,401}} = \frac{2}{7}$, and because:

$$(0.1)^5 = 0.00001$$

$$\sqrt[5]{0.00001} = 0.1$$

Arithmetic Concepts

Estimation

Sometimes it is helpful to find an estimated answer to a problem rather than working out an exact answer. An **estimation** might be much quicker to find, and given the scenario, an estimation might be all that is required. For example, if Aria goes grocery shopping and has only a $100 bill to cover all of her purchases, it might be appropriate for her to estimate the total of the items she is purchasing to determine if she has enough money to cover them. Also, an estimation can help determine if an answer makes sense. For instance, if an answer in the 100s is expected, but the result is a fraction less than 1, something is probably wrong in the calculation.

The first type of estimation involves rounding. **Rounding** consists of expressing a number in terms of the nearest decimal place like the tenth, hundredth, or thousandth place, or in terms of the nearest whole number unit like tens, hundreds, or thousands place. When rounding to a specific place value, look at the digit to the right of the place. If it is 5 or higher, round the number to its left up to the next value, and if it is 4 or lower, keep that number at the same value. For instance, 1,654.2674 rounded to the nearest thousand is 2,000, and the same number rounded to the nearest thousandth is 1,654.267. Rounding can be used in the scenario when grocery totals need to be estimated. Items can be rounded to the nearest dollar. For example, a can of corn that costs $0.79 can be rounded to $1.00, and then all other items can be rounded in a similar manner and added together.

When working with larger numbers, it might make more sense to round to higher place values. For example, when estimating the total value of a dealership's car inventory, it would make sense to round the car values to the nearest thousands place. The price of a car that is on sale for $15,654 can be estimated at $16,000. All other cars on the lot could be rounded in the same manner, and then their sum can be found. Depending on the situation, it might make sense to calculate an over-estimate.

For example, to make sure Aria has enough money at the grocery store, rounding up every time for each item would ensure that she will have enough money when it comes time to pay. A $0.40 item rounded up to $1.00 would ensure that there is a dollar to cover that item. Traditional rounding rules would round $0.40 to $0, which does not make sense in this particular real-world setting. Aria might not have a dollar available at checkout to pay for that item if she uses traditional rounding. It is up to the customer to decide the best approach when estimating.

Estimating is also very helpful when working with measurements. Bryan is updating his kitchen and wants to retile the floor. Again, an over-measurement might be useful. Also, rounding to nearest half-unit might be helpful. For instance, one side of the kitchen might have an exact measurement of 14.32 feet, and the most useful measurement needed to buy tile could be estimating this quantity to be 14.5 feet. If the kitchen was rectangular and the other side measured 10.9 feet, Bryan might round the other side to 11 feet. Therefore, Bryan would find the total tile necessary according to the following area calculation:

$$14.5 \times 11 = 159.5 \text{ square feet}$$

To make sure he purchases enough tile, Bryan would probably want to purchase at least 160 square feet of tile. This is a scenario in which an estimation might be more useful than an exact calculation. Having more tile than necessary is better than having an exact amount, in case any tiles are broken or otherwise unusable.

Finally, estimation is helpful when exact answers are necessary. Consider a situation in which Sabina has many operations to perform on numbers with decimals, and she is allowed a calculator to find the result. Even though an exact result can be obtained with a calculator, there is always a possibility that Sabina could make an error while inputting the data.

For example, she could miss a decimal place, or misuse a parenthesis, causing a problem with the actual order of operations. In this case, a quick estimation at the beginning would be helpful to make sure the final answer is given with the correct number of units. Sabina has to find the exact total of 10 cars listed for sale at the dealership. Each price has two decimal places included to account for both dollars and cents. If one car is listed at $21,234.43 but Sabina incorrectly inputs into the calculator the price of $2,123.443, this error would throw off the final sum by almost $20,000. A quick estimation at the beginning, by rounding each price to the nearest thousands place and finding the sum of the prices, would give Sabina an amount to compare the exact amount to. This comparison would let Sabina see if an error was made in her exact calculation.

Percent

Percentages are defined to be parts per one hundred. To convert a decimal to a percentage, move the decimal point two units to the right and place the percent sign after the number. Percentages appear in many scenarios in the real world. It is important to make sure the statement containing the percentage is translated to a correct mathematical expression. Be aware that it is extremely common to make a mistake when working with percentages within word problems.

An example of a word problem containing a percentage is the following: 35% of people speed when driving to work. In a group of 5,600 commuters, how many would be expected to speed on the way to their place of employment? The answer to this problem is found by finding 35% of 5,600. First, change the percentage to the decimal 0.35. Then compute the product: $0.35 \times 5{,}600 = 1{,}960$. Therefore, it would be expected that 1,960 of those commuters would speed on their way to work based on the data given. In this situation, the word "of" signals to use multiplication to find the answer.

Another way percentages are used is in the following problem: Teachers work 8 months out of the year. What percent of the year do they work? To answer this problem, find what percent of 12 the number 8 is, because there are 12 months in a year. Therefore, divide 8 by 12, and convert that number to a percentage:

$$\frac{8}{12} = \frac{2}{3} = 0.66\overline{6}$$

The percentage rounded to the nearest tenth place tells us that teachers work 66.7% of the year. Percentages also appear in real-world application problems involving finding missing quantities like in the following question: 60% of what number is 75? To find the missing quantity, an equation can be used. Let x be equal to the missing quantity. Therefore:

$$0.60x = 75$$

Divide each side by 0.60 to obtain 125. Therefore, 60% of 125 is equal to 75.

Sales tax is an important application relating to percentages because tax rates are usually given as percentages. For example, a city might have an 8% sales tax rate. Therefore, when an item is purchased with that tax rate, the real cost to the customer is 1.08 times the price in the store. For example, a $25 pair of jeans costs the customer:

$$\$25 \times 1.08 = \$27$$

Sales tax rates can also be determined if they are unknown when an item is purchased. If a customer visits a store and purchases an item for $21.44, but the price in the store was $19, they can find the tax rate by first subtracting $\$21.44 - \$19$ to obtain $2.44, the sales tax amount. The sales tax is a percentage of the in-store price. Therefore, the tax rate is $\frac{2.44}{19} = 0.128$, which has been rounded to the nearest thousandths place. In this scenario, the actual sales tax rate given as a percentage is 12.8%.

Ratio/Rate

Fractions appear in everyday situations, and in many scenarios, they appear in the real-world as ratios and in proportions. A **ratio** is formed when two different quantities are compared. For example, in a group of 50 people, if there are 33 females and 17 males, the ratio of females to males is 33 to 17. This expression can be written in the fraction form as $\frac{33}{50}$, where the denominator is the sum of females and males, or by using the ratio symbol, 33:17. The order of the number matters when forming ratios. In the same setting, the ratio of males to females is 17 to 33, which is equivalent to $\frac{17}{50}$ or 17:33. A **proportion** is an equation involving two ratios.

The equation $\frac{a}{b} = \frac{c}{d}$, or $a:b = c:d$ is a proportion, for real numbers a, b, c, and d. Usually, in one ratio, one of the quantities is unknown, and cross-multiplication is used to solve for the unknown. Consider $\frac{1}{4} = \frac{x}{5}$. To solve for x, cross-multiply to obtain $5 = 4x$. Divide each side by 4 to obtain the solution $x = \frac{5}{4}$. It is also true that percentages are ratios in which the second term is 100 minus the first term. For example, 65% is 65:35 or $\frac{65}{100}$. Therefore, when working with percentages, one is also working with ratios.

Real-world problems frequently involve proportions. For example, consider the following problem: If 2 out of 50 pizzas are usually delivered late from a local Italian restaurant, how many would be late out of 235 orders? The following proportion would be solved with x as the unknown quantity of late pizzas:

$$\frac{2}{50} = \frac{x}{235}$$

Cross multiplying results in $470 = 50x$. Divide both sides by 50 to obtain $x = \frac{470}{50}$, which in lowest terms is equal to $\frac{47}{5}$. In decimal form, this improper fraction is equal to 9.4. Because it does not make sense to answer this question with decimals (portions of pizzas do not get delivered) the answer must be rounded.

Traditional rounding rules would say that 9 pizzas would be expected to be delivered late. However, to be safe, rounding up to 10 pizzas out of 235 would probably make more sense.

Recall that a ratio is the comparison of two different quantities. Comparing 2 apples to 3 oranges results in the ratio 2:3, which can be expressed as the fraction $\frac{2}{5}$. Note that order is important when discussing ratios. The number mentioned first is the antecedent, and the number mentioned second is the consequent. Note that the consequent of the ratio and the denominator of the fraction are *not* the same. When there are 2 apples to 3 oranges, there are five fruit total; two fifths of the fruit are apples, while three fifths are oranges. The ratio 2:3 represents a different relationship than the ratio 3:2. Also, it is important to make sure that when discussing ratios that have units attached to them, the two quantities use the same units. For example, to think of 8 feet to 4 yards, it would make sense to convert 4 yards to feet by multiplying by 3. Therefore, the ratio would be 8 feet to 12 feet, which can be expressed as the fraction $\frac{8}{20}$. Also, note that it is proper to refer to ratios in lowest terms. Therefore, the ratio of 8 feet to 4 yards is equivalent to the fraction $\frac{2}{5}$.

Many real-world problems involve ratios. Often, problems with ratios involve proportions, as when two ratios are set equal to find the missing amount. However, some problems involve deciphering single ratios. For example, consider an amusement park that sold 345 tickets last Saturday. If 145 tickets were sold to adults and the rest of the tickets were sold to children, what would the ratio of the number of adult tickets to children's tickets be? A common mistake would be to say the ratio is 145:345. However, 345 is the total number of tickets sold, not the number of children's tickets. There were $345 - 145 = 200$ tickets sold to children. The correct ratio of adult to children's tickets is 145:200. As a fraction, this expression is written as $\frac{145}{345}$, which can be reduced to $\frac{29}{69}$.

While a ratio compares two measurements using the same units, rates compare two measurements with different units. Examples of rates would be \$200 for 8 hours of work, or 500 miles traveled per 20 gallons. Because the units are different, it is important to always include the units when discussing rates. Rates can be easily seen because if they are expressed in words, the two quantities are usually split up using one of the following words: *for, per, on, from, in*. Just as with ratios, it is important to write rates in lowest terms. A common rate that can be found in many real-life situations is cost per unit. This quantity describes how much one item or one unit costs. This rate allows the best buy to be determined, given a couple of different sizes of an item with different costs. For example, if 2 quarts of soup was sold for \$3.50 and 3 quarts was sold for \$4.60, to determine the best buy, the cost per quart should be found. $\frac{\$3.50}{2\text{ qt}} = \$1.75$ per quart, and $\frac{\$4.60}{3\text{ qt}} = \$1.53$ per quart. Therefore, the better deal would be the 3-quart option.

Rate of change problems involve calculating a quantity per some unit of measurement. Usually the unit of measurement is time. For example, meters per second is a common rate of change. To calculate this measurement, find the amount traveled in meters and divide by total time traveled. The calculation is an average of the speed over the entire time interval. Another common rate of change used in the real world is miles per hour. Consider the following problem that involves calculating an average rate of change in temperature. Last Saturday, the temperature at 1:00 a.m. was 34 degrees Fahrenheit, and at noon, the temperature had increased to 75 degrees Fahrenheit. What was the average rate of change over that time interval? The average rate of change is calculated by finding the change in temperature and dividing by the total hours elapsed. Therefore, the rate of change was equal to $\frac{75-34}{12-1} = \frac{41}{11}$ degrees per hour. This quantity rounded to two decimal places is equal to 3.72 degrees per hour.

A common rate of change that appears in algebra is the slope calculation. Given a linear equation in one variable, $y = mx + b$, the **slope**, m, is equal to $\frac{rise}{run}$ or $\frac{change\ in\ y}{change\ in\ x}$. In other words, slope is equivalent to the ratio of the vertical and horizontal changes between any two points on a line. The vertical change is known as the **rise**, and the horizontal change is known as the **run**. Given any two points on a line (x_1, y_1) and (x_2, y_2), slope can be calculated with the formula:

$$m = \frac{y_2 - y_1}{x_2 - x_1} = \frac{\Delta y}{\Delta x}$$

Common real-world applications of slope include determining how steep a staircase should be, calculating how steep a road is, and determining how to build a wheelchair ramp.

Many times, problems involving rates and ratios involve proportions. A proportion states that two ratios (or rates) are equal. The property of cross products can be used to determine if a proportion is true, meaning both ratios are equivalent. If $\frac{a}{b} = \frac{c}{d'}$, then to clear the fractions, multiply both sides by the least common denominator, bd. This results in $ad = bc$, which is equal to the result of multiplying along both diagonals. For example, $\frac{4}{40} = \frac{1}{10}$ grants the cross product $4 \times 10 = 40 \times 1$, which is equivalent to $40 = 40$ and shows that this proportion is true. Cross products are used when proportions are involved in real-world problems. Consider the following: If 3 pounds of fertilizer will cover 75 square feet of grass, how many pounds are needed for 375 square feet? To solve this problem, a proportion can be set up using two ratios. Let x equal the unknown quantity, pounds needed for 375 feet. Then, the equation found by setting the two given ratios equal to one another is:

$$\frac{3}{75} = \frac{x}{375}$$

Cross-multiplication gives $3 \times 375 = 75x$. Therefore, $1{,}125 = 75x$. Divide both sides by 75 to get $x = 15$. Therefore, 15 pounds of fertilizer are needed to cover 375 square feet of grass.

Absolute Value

The **absolute value** of any real number is the distance from that number to 0 on the number line. The absolute value of a number can never be negative. For example, the absolute value of both 8 and −8 is 8 because they are both 8 units away from 0 on the number line. This is written as:

$$|8| = |-8| = 8$$

The Number Line

A **number line** is a great way to visualize the integers. Integers are labeled on the following number line:

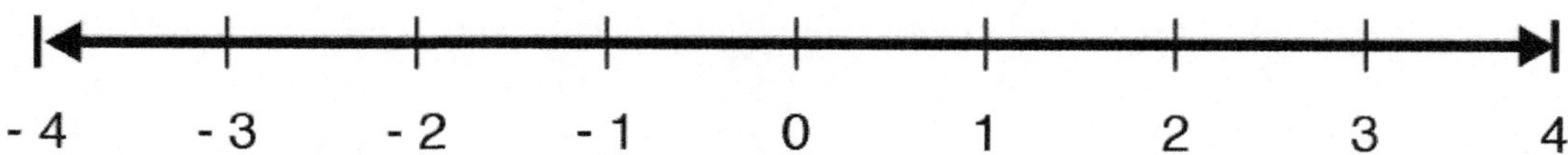

The arrows on the right- and left-hand sides of the number line show that the line continues indefinitely in both directions.

The number line contains all real numbers. To graph a number other than an integer on a number line, it needs to be plotted between two integers. For example, 3.5 would be plotted halfway between 3 and 4.

Decimal Representation

Fractions can be converted to decimals. With a calculator, a fraction is converted to a decimal by dividing the numerator by the denominator. For example:

$$\frac{2}{5} = 2 \div 5 = 0.4$$

Sometimes, rounding might be necessary. Consider:

$$\frac{2}{7} = 2 \div 7 = 0.28571429$$

This decimal could be rounded for ease of use, and if it needed to be rounded to the nearest thousandth, the result would be 0.286. If a calculator is not available, a fraction can be converted to a decimal manually. First, find a number that, when multiplied by the denominator, has a value equal to 10, 100, 1,000, etc. Then, multiply both the numerator and denominator times that number. The decimal form of the fraction is equal to the new numerator with a decimal point placed as many place values to the left as there are zeros in the denominator. For example, to convert $\frac{3}{5}$ to a decimal, multiply both the numerator and denominator times 2, which results in $\frac{6}{10}$.

The decimal is equal to 0.6 because there is one zero in the denominator, and so the decimal place in the numerator is moved one unit to the left. In the case where rounding would be necessary while working without a calculator, an approximation must be found. A number close to 10, 100, 1,000, etc. can be used. For example, to convert $\frac{1}{3}$ to a decimal, the numerator and denominator can be multiplied by 33 to turn the denominator into approximately 100, which makes for an easier conversion to the equivalent decimal. This process results in $\frac{33}{99}$ and an approximate decimal of 0.33. Once in decimal form, the number can be converted to a percentage. Multiply the decimal by 100 and then place a percent sign after the number. For example, 0.614 is equal to 61.4%. In other words, move the decimal place two units to the right and add the percentage symbol.

Sequences of Numbers

A **sequence of numbers** is a list of numbers that follows a specific pattern. Each member of the sequence is known as an **individual term** of the sequence, and a formula can be found to represent each term. For example, the list of numbers 5, 10, 15, 20, ... is a sequence of numbers, and ... shows that the sequence continues indefinitely. Each term represents a multiple of 5. The first term is 1×5, the second term is 2×5, the third term is 3×5, etc. In general, the nth term is $5 \times n$.

Other, more complicated sequences can exist as well. For example, each term of the sequence 1, 4, 9, 16, 25, ... is not found through addition or multiplication. Each term happens to be a perfect square, and the first term is 1 squared, the second term is 2 squared, etc. Therefore, the nth term is n^2. A famous sequence of numbers is the **Fibonacci sequence**, which consists of 0, 1, 1, 2, 3, 5, 8, 13, 21, ... After the first two terms 0 and 1, all terms are found by adding the two previous terms. Therefore, the next term in the sequence is:

$$13 + 21 = 34$$

The formula for the nth term is defined recursively, using the two previous terms.

Algebra

Operations with Exponents

When performing operations with exponents, one must make sure that the **order of operations** is followed. Therefore, once all operations within any parentheses or grouping symbols are performed, all exponents must be evaluated before any other operation is completed. For example, to evaluate:

$$(3-5)^2+4-5^3$$

the subtraction in parentheses is done first. Then the exponents are evaluated to obtain:

$$(-2)^2+4-5^3, \text{ or } 4+4-125$$

which is equivalent to -117.

A common mistake involves negative signs combined with exponents. Note that -4^2 is not equal to $(-4)^2$. The negative sign is in front of the first exponential expression, so the result is:

$$-(4\times 4)=-16$$

However, the negative sign is inside the parentheses in the second expression, so the result is :

$$(-4\times -4)=16$$

Laws of exponents can also help when performing operations with exponents. If two exponential expressions have the same base and are being multiplied, just add the exponents. For example:

$$2^5\times 2^7=2^{5+7}=2^{12}$$

If two exponential expressions have the same base and are being divided, subtract the exponents. For example:

$$\frac{4^{30}}{4^2}=4^{30-2}=4^{28}$$

If an exponential expression is being raised to another exponent, multiply the exponents together. For example:

$$(3^2)^5=3^{2\times 5}=3^{10}$$

Factoring and Simplifying Algebraic Expressions

A factorization of an algebraic expression can be found. Throughout the process, a more complicated expression can be decomposed into products of simpler expressions. To factor a polynomial, first determine if there is a greatest common factor. If there is, factor it out. For example:

$$2x^2+8x$$

has a greatest common factor of $2x$ and can be written as:

$$2x(x+4)$$

Once the greatest common monomial factor is factored out, if applicable, count the number of terms in the polynomial. If there are two terms, is it a difference of squares, a sum of cubes, or a difference of cubes? If so, the following rules can be used:

$$a^2 - b^2 = (a + b)(a - b)$$

$$a^3 + b^3 = (a + b)(a^2 - ab + b^2)$$

$$a^3 - b^3 = (a - b)(a^2 + ab + b^2)$$

If there are three terms, and if the trinomial is a perfect square trinomial, it can be factored into the following:

$$a^2 + 2ab + b^2 = (a + b)^2$$

$$a^2 - 2ab + b^2 = (a - b)^2$$

If not, try factoring into a product of two binomials by trial and error into a form of $(x + p)(x + q)$. For example, to factor:

$$x^2 + 6x + 8$$

determine what two numbers have a product of 8 and a sum of 6. Those numbers are 4 and 2, so the trinomial factors into:

$$(x + 2)(x + 4)$$

Finally, if there are four terms, try factoring by grouping. First, group terms together that have a common monomial factor. Then, factor out the common monomial factor from the first two terms. Next, look to see if a common factor can be factored out of the second set of two terms that results in a common binomial factor.

Finally, factor out the common binomial factor of each expression, for example:

$$xy - x + 5y - 5$$

$$x(y - 1) + 5(y - 1)$$

$$(y - 1)(x + 5)$$

After the expression is completely factored, check to see if the factorization is correct by multiplying to try to obtain the original expression. Factorizations are helpful in solving equations that consist of a polynomial set equal to zero. If the product of two algebraic expressions equals zero, then at least one of the factors is equal to zero. Therefore, factor the polynomial within the equation, set each factor equal to zero, and solve. For example, $x^2 + 7x - 18 = 0$ can be solved by factoring into:

$$(x + 9)(x - 2) = 0$$

Set each factor equal to zero and solve to obtain $x = -9$ and $x = 2$.

Relations, Functions, Equations, and Inequalities

In math, a **relation** is a relationship between two sets of numbers. By using a rule, it takes a number from the first set and matches it to a number in the second set. A relation consists of a set of inputs, known as the **domain**, and a set of outputs, known as the **range**. A **function** is a relation in which each member of the domain is paired to only one other member of the range. In other words, each input has only one output.

Here is an example of a relation that is not a function:

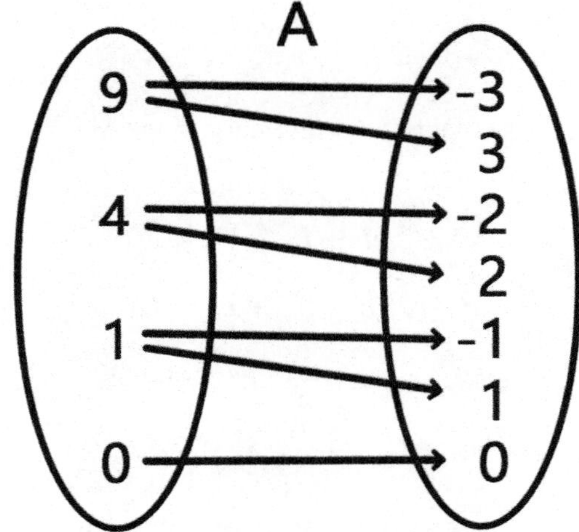

Every member of the first set, the domain, is mapped to two members of the second set, the range. Therefore, this relation is not a function.

In addition to a diagram representing sets, a function can be represented by a table of **ordered pairs**, a graph of ordered pairs (**a scatterplot**), or a set of ordered pairs as shown in the following:

Mapping

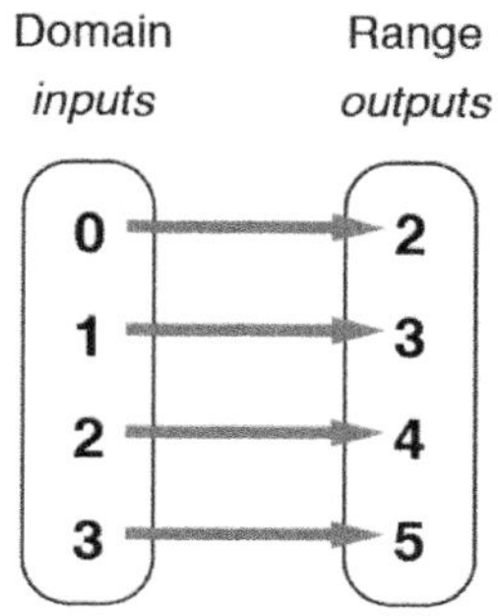

Domain *inputs*		Range *outputs*
0	→	2
1	→	3
2	→	4
3	→	5

Table

x	y
0	2
1	3
2	4
3	5

Graph

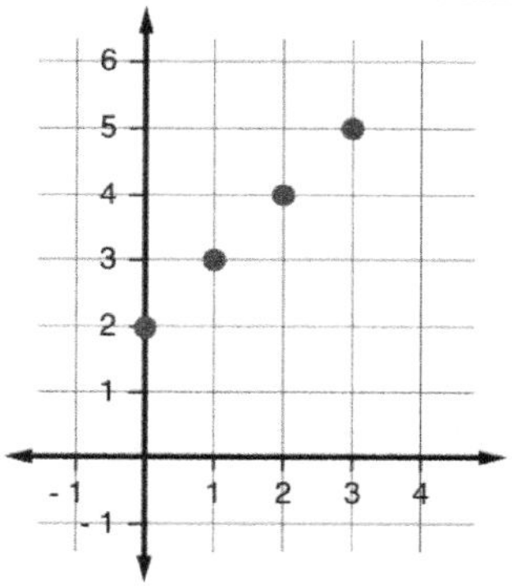

Ordered Pairs

{(0,2),(1,3),(2,4),(3,5)}

Note that this relation is a function because every member of the domain is mapped to exactly one member of the range.

An **equation** occurs when two algebraic expressions are set equal to one another. Functions can be represented in equation form. Given an equation in two variables, *x* and *y*, it can be expressed in function form if solved for *y*. For example, the linear equation $2x + y = 5$ can be solved for *y* to obtain:

$$y = -2x + 5$$

otherwise known as **slope-intercept** form. To place the equation in function form, replace *y* with *f(x)*, which is read "*f* of *x*." Therefore:

$$f(x) = -2x + 5$$

This notation clarifies the input–output relationship of the function. The function *f* is a function of *x*, so an *x* value can be plugged into the function to obtain an output. For example:

$$f(2) = -2 \times 2 + 5 = 1$$

Therefore, an input of 2 corresponds to an output of 1.

A function can be graphed by plotting ordered pairs in the *xy*-plane in the same way that the equation form is graphed. The graph of a function always passes the **Vertical Line Test**. Basically, for any graph, if a vertical line can be drawn through any part of the graph and it hits the graph in more than one place, the graph is not a function. For example, the graph of a circle is not a function. The Vertical Line Test shows that with these relationships, the same *x* value has more than one *y* value, which goes against the definition of a function.

Inequalities look like equations, but instead of an equals sign, $<, >, \leq, \geq$, or $\neq$ are used. Here are some examples of inequalities:

$$2x + 7 < y, 3x^2 \geq 5$$

and

$$x \neq 4$$

Inequalities show relationships between algebraic expressions when the quantities are different. Inequalities can also be expressed in function form if they are solved for *y*. For instance, the first inequality listed above can be written as:

$$2x + y < f(x)$$

Solving Equations

Solving Linear Equations

An **equation in one variable** is a mathematical statement where two algebraic expressions in one variable, usually x, are set equal. To solve the equation, the variable must be isolated on one side of the equals sign. The addition and multiplication principles of equality are used to isolate the variable. The **addition principle of equality** states that the same number can be added to or subtracted from both sides of an equation. Because the same value is being used on both sides of the equals sign, equality is maintained. For example, the equation:

$$2x = 5x$$

is equivalent to both:

$$2x + 3 = 5x + 3 \; and \; 2x - 5 = 5x - 5$$

This principle can be used to solve the following equation:

$$x + 5 = 4$$

The variable x must be isolated, so to move the 5 from the left side, subtract 5 from both sides of the equals sign. Therefore:

$$x + 5 - 5 = 4 - 5$$

So, the solution is $x = -1$.

This process illustrates the idea of an **additive inverse** because subtracting 5 is the same as adding -5. Basically, add the opposite of the number that must be removed to both sides of the equals sign. The **multiplication principle of equality** states that equality is maintained when a number is either multiplied times both expressions on each side of the equals sign, or when both expressions are divided by the same number.

For example, $4x = 5$ is equivalent to both $16x = 20$ and $x = \frac{5}{4}$.

Multiplying both sides times 4 and dividing both sides by 4 maintains equality. Solving the equation $6x - 18 = 5$ requires the use of both principles.

First, apply the addition principle to add 18 to both sides of the equals sign, which results in $6x = 23$.

Then use the multiplication principle to divide both sides by 6, giving the solution $x = \frac{23}{6}$.

Using the multiplication principle in the solving process is the same as involving a multiplicative inverse. A **multiplicative inverse** is a value that, when multiplied by a given number, results in 1. Dividing by 6 is the same as multiplying by $\frac{1}{6}$, which is both the reciprocal and multiplicative inverse of 6.

When solving a linear equation in one variable, checking the answer shows if the solution process was performed correctly. Plug the solution into the variable in the original equation. If the result is a false statement, something was done incorrectly during the solution procedure. Checking the example above gives the following:

$$6 \times \frac{23}{6} - 18 = 23 - 18 = 5$$

Therefore, the solution is correct.

Some equations in one variable involve fractions or the use of the distributive property. In either case, the goal is to obtain only one variable term and then use the addition and multiplication principles to isolate that variable. Consider the equation:

$$\frac{2}{3}x = 6$$

To solve for x, multiply each side of the equation by the reciprocal of $\frac{2}{3}$, which is $\frac{3}{2}$. This step results in:

$$\frac{3}{2} \times \frac{2}{3}x = \frac{3}{2} \times 6$$

which simplifies into the solution $x = 9$. Now consider the equation:

$$3(x + 2) - 5x = 4x + 1$$

Use the distributive property to clear the parentheses. Therefore, multiply each term inside the parentheses by 3. This step results in:

$$3x + 6 - 5x = 4x + 1$$

Next, collect like terms on the left-hand side. **Like terms** are terms with the same variable or variables raised to the same exponent(s). Only like terms can be combined through addition or subtraction. After collecting like terms, the equation is:

$$-2x + 6 = 4x + 1$$

Finally, apply the addition and multiplication principles.

Add $2x$ to both sides to obtain $6 = 6x + 1$.

Then, subtract 1 from both sides to obtain $5 = 6x$. Finally, divide both sides by 6 to obtain the solution $\frac{5}{6} = x$.

Two other types of solutions can be obtained when solving an equation in one variable. The final result could be that there is either no solution or that the solution set contains all real numbers. Consider the equation:

$$4x = 6x + 5 - 2x$$

First, the like terms can be combined on the right to obtain:

$$4x = 4x + 5$$

Next, subtract $4x$ from both sides. This step results in the false statement $0 = 5$. There is no value that can be plugged into x that will ever make this equation true. Therefore, there is no solution. The solution procedure contained correct steps, but the result of a false statement means that no value satisfies the equation. The symbolic way to denote that no solution exists is $\emptyset$. Next, consider the equation:

$$5x + 4 + 2x = 9 + 7x - 5$$

Combining the like terms on both sides results in:

$$7x + 4 = 7x + 4$$

The left-hand side is exactly the same as the right-hand side. Using the addition principle to move terms, the result is $0 = 0$, which is always true. Therefore, the original equation is true for any number, and the solution set is all real numbers. The symbolic way to denote such a solution set is $\mathbb{R}$, or in interval notation, $(-\infty, \infty)$.

One-step problems take only one mathematical step to solve. For example, solving the equation $5x = 45$ is a one-step problem because the one step of dividing both sides of the equation by 5 is the only step necessary to obtain the solution $x = 9$. The multiplication principle of equality is the one step used to isolate the variable. The equation is of the form $ax = b$, where a and b are rational numbers. Similarly, the addition principle of equality could be the one step needed to solve a problem. In this case, the equation would be of the form $x + a = b$ or $x - a = b$, for real numbers a and b.

A **multi-step problem** involves more than one step to find the solution, or it could consist of solving more than one equation. An equation that involves both the addition principle and the multiplication principle is a two-step problem, and an example of such an equation is $2x - 4 = 5$. Solving involves adding 4 to both sides and then dividing both sides by 2. An example of a two-step problem involving two separate equations is:

$$y = 3x \text{ and } 2x + y = 4$$

The two equations form a system of two equations that must be solved together in two variables. The system can be solved by the substitution method. Since y is already solved for in terms of x, plug $3x$ in for y into the equation $2x + y = 4$, resulting in:

$$2x + 3x = 4$$

Therefore, $5x = 4$ and $x = \frac{4}{5}$. Because there are two variables, the solution consists of a value for both x and for y. Substitute $x = \frac{4}{5}$ into either original equation to find y. The easiest choice is $y = 3x$. Therefore:

$$y = 3 \times \frac{4}{5} = \frac{12}{5}$$

The solution can be written as the ordered pair $\left(\frac{4}{5}, \frac{12}{5}\right)$.

Real-world problems can be translated into both one-step and multi-step problems. In either case, the word problem must be translated from the verbal form into mathematical expressions and equations that can be solved using algebra.

An example of a one-step real-world problem is the following: A cat weighs half as much as a dog living in the same house. If the dog weighs 14.5 pounds, how much does the cat weigh? To solve this problem, an equation can be used. In any word problem, the first step must be defining variables that represent the unknown quantities. For this problem, let x be equal to the unknown weight of the cat. Because two times the weight of the cat equals 14.5 pounds, the equation to be solved is: $2x = 14.5$. Use the multiplication principle to divide both sides by 2. Therefore, $x = 7.25$. The cat weighs 7.25 pounds.

Most of the time, real-world problems are more difficult than this one and consist of multi-step problems. The following is an example of a multi-step problem: The sum of two consecutive page numbers is equal to 437. What are those page numbers?

First, define the unknown quantities. If x is equal to the first page number, then $x + 1$ is equal to the next page number because they are consecutive integers. Their sum is equal to 437, and this statement translates to the equation:

$$x + x + 1 = 437$$

To solve, first collect like terms to obtain:

$$2x + 1 = 437$$

Then, subtract 1 from both sides and then divide by 2. The solution to the equation is $x = 218$. Therefore, the two consecutive page numbers that satisfy the problem are 218 and 219.

It is always important to make sure that answers to real-world problems make sense. For instance, it should be a red flag if the solution to this same problem resulted in decimals, which would indicate the need to check the work. Page numbers are whole numbers; therefore, if decimals are found to be answers, the solution process should be double-checked to see where mistakes were made.

Solving Quadratic Equations

A **quadratic equation** in standard form:

$$ax^2 + bx + c = 0$$

can have either two solutions, one solution, or two complex solutions (no real solutions). This is determined using the determinant:

$$b^2 - 4ac$$

If the determinant is positive, there are two real solutions. If the determinant is negative, there are no real solutions. If the determinant is equal to 0, there is one real solution. For example, given the quadratic equation $4x^2 - 2x + 1 = 0$, its determinant is:

$$(-2)^2 - 4(4)(1) = 4 - 16 = -12$$

so it has two complex solutions, meaning no real solutions.

There are quite a few ways to solve a quadratic equation. The first is by **factoring**. If the equation is in standard form and the polynomial can be factored, set each factor equal to 0, and solve using the Principle of Zero Products. For example:

$$x^2 - 4x + 3 = (x - 3)(x - 1)$$

Therefore, the solutions of $x^2 - 4x + 3 = 0$ are those that satisfy both $x - 3 = 0$ and $x - 1 = 0$, or $x = 3$ and $x = 1$.

This is the simplest method to solve quadratic equations; however, not all polynomials inside the quadratic equations can be factored.

Another method is **completing the square**. The polynomial $x^2 + 10x - 9$ cannot be factored, so the next option is to complete the square in the equation:

$$x^2 + 10x - 9 = 0$$

to find its solutions. The first step is to add 9 to both sides, moving the constant over to the right side, resulting in:

$$x^2 + 10x = 9$$

Then the coefficient of x is divided by 2 and squared. This result is then added to both sides of the equation. In this example, $\left(\frac{10}{2}\right)^2 = 25$ is added to both sides of the equation to obtain:

$$x^2 + 10x + 25 = 9 + 25 = 34$$

The left-hand side can then be factored into:

$$(x + 5)^2 = 34$$

Solving for x then involves taking the square root of both sides and subtracting 5. This leads to the two solutions:

$$x = \pm\sqrt{34} - 5$$

The third method is the **quadratic formula**. Given a quadratic equation in standard form, $ax^2 + bx + c = 0$, its solutions always can be found using the formula:

$$x = \frac{-b \pm \sqrt{b^2 - 4ac}}{2a}$$

Solving Inequalities

Inequalities can be solved in a similar method as equations. Basically, the goal is to isolate the variable, and this process can be completed by adding numbers onto both sides, subtracting numbers off of both sides, multiplying numbers onto both sides, and dividing numbers off of both sides of the inequality. Basically, if something is done to one side, it has to be done to the other side, just like when solving equations. However, there is one important difference, and that difference occurs when multiplying times negative numbers and dividing by negative numbers. If either one of these steps must be performed in the solution process, the inequality symbol must be reversed.

Consider the following inequality:

$$2 - 3x < 11$$

The goal is to isolate the variable *x*, so first subtract 2 off both sides to obtain $-3x < 9$. Then divide both sides by –3, making sure to "*flip the sign*." This results in $x > -3$, which is the solution set. This solution set means that all numbers greater than –3 satisfy the original inequality, and therefore any number larger than –3 is a solution. In **set-builder notation**, this set can be written as $\{x|x > -3\}$, which is read "all *x* values such that *x* is greater than –3." In addition to the inequality form of the solution, solutions of inequalities can be expressed by using both a number line and interval notation.

Here is a chart that highlights all three types of expressing the solutions:

Interval Notation	Number Line Sketch	Set-builder Notation
(a , b)	open circle at a, open circle at b, segment between	{ x I a < x < b}
(a , b]	open circle at a, closed circle at b, segment between	{ x I a < x ≤ b}
[a , b)	closed circle at a, open circle at b, segment between	{ x I a ≤ x < b}
[a , b]	closed circle at a, closed circle at b, segment between	{ x I a ≤ x ≤ b}
(a , ∞)	open circle at a, ray to the right	{ x I x > a}
(- ∞ , b)	ray from the left, open circle at b	{ x I x < b}
[a , ∞)	closed circle at a, ray to the right	{ x I x ≥ a}
(- ∞ , b]	ray from the left, closed circle at b	{ x I x ≤ b}
(- ∞, ∞)	entire line	ℝ

Solving Simultaneous Equations

Simultaneous equations, otherwise known as a **system of equations**, can be solved in a variety of ways. If the system consists of two linear equations, they can be solved by graphing, substitution, or elimination.

When **graphing**, the solution of a system of equations is the point of intersection. If the two lines are parallel, they will never intersect, and there is no solution. If the two lines are the same, there are infinitely many solutions, and the solution set is equal to the entire line.

Here are the three cases:

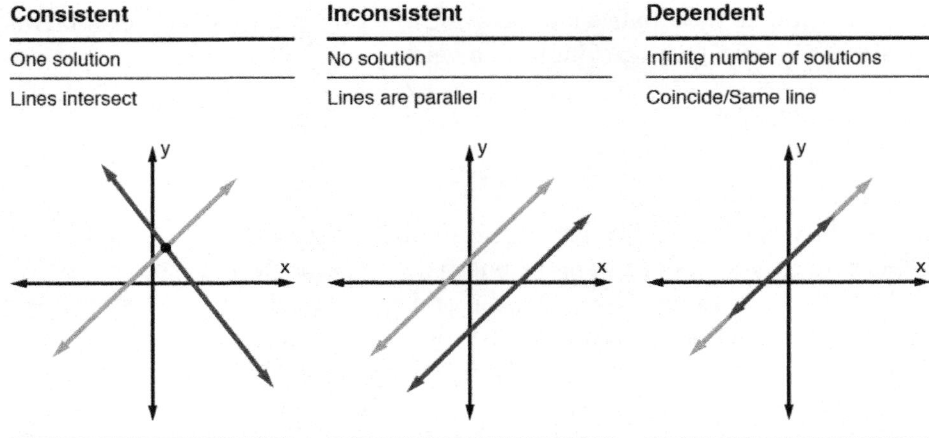

There are two algebraic methods to finding solutions. The first is **substitution**. This is better suited when one of the equations is already solved for one variable, or it is easy to do so. Then this equation gets substituted into the other equation for that variable, resulting in an equation in one variable. Solve for the given variable and plug that value into one of the original equations to find the other variable. This last step is known as **back-substitution**.

Here is an example of solving a system with the substitution method:

$$y = x + 1 \qquad 2y = 3x$$

$$2y = 3x$$

$$2(x + 1) = 3x$$

$$2x + 2 = 3x$$

$$-2x \qquad -2x$$

$$2 = x$$

$$y = x + 1$$

$$y = 2 + 1 = 3$$

Solution: (2 , 3)

The other method is known as **elimination**, or the **addition method**. This is better suited when the equations are in standard form:

$$Ax + By = C$$

The goal in this method is to multiply one or both equations times numbers that result in opposite coefficients. Then add the equations together to obtain an equation in one variable. Solve for the given variable, and then take that value and back-substitute to obtain the other part of the ordered pair solution. Here is an example of elimination:

$$\begin{cases} -x + 5y = 8 \\ 3x + 7y = -2 \end{cases} \overset{\times 3}{\Longrightarrow} \begin{cases} -3x + 15y = 24 \\ 3x + 7y = -2 \end{cases}$$

$$\begin{array}{r} -3x + 15y = 24 \\ 3x + 7y = -2 \\ \hline 22y = 22 \end{array}$$

$$\frac{22y}{22} = \frac{22}{22}$$

$$y = 1$$

Note that in order to check an answer when solving a system of equations, the solution must be checked in both original equations to show that it solves both equations.

Setting Up Equations

When presented with a real-world problem that must be solved, the first step is always to determine what the unknown quantity is that must be solved for. Use a variable, such as x or t, to represent that unknown quantity. Sometimes there can be two or more unknown quantities. In this case, either choose an additional variable, or if a relationship exists between the unknown quantities, express the other quantities in terms of the original variable. After choosing the variables, form algebraic expressions and/or equations that represent the verbal statement in the problem. The following table shows examples of vocabulary used to represent the different operations.

Addition	Sum, plus, total, increase, more than, combined, in all
Subtraction	Difference, less than, subtract, reduce, decrease, fewer, remain
Multiplication	Product, multiply, times, part of, twice, triple
Division	Quotient, divide, split, each, equal parts, per, average, shared

The combination of operations and variables form both mathematical expression and equations. The difference between expressions and equations is that there is no equals sign in an expression, and that expressions are **evaluated** to find an unknown quantity, while equations are **solved** to find an unknown quantity. Also, inequalities can exist within verbal mathematical statements. Instead of a statement of equality, expressions state quantities are *less than*, *less than or equal to*, *greater than*, or *greater than or equal to*. Another type of inequality is when a quantity is said to be *not equal to* another quantity. The symbol used to represent "not equal to" is $\neq$.

The steps for solving inequalities in one variable are the same steps for solving equations in one variable. The addition and multiplication principles are used. However, to maintain a true statement when using the $<, \leq, >$, and $\geq$ symbols, if a negative number is either multiplied times both sides of an inequality or divided from both sides of an inequality, the sign must be flipped. For instance, consider the following

inequality: $3 - 5x \leq 8$. First, 3 is subtracted from each side to obtain $-5x \leq 5$. Then, both sides are divided by -5, while flipping the sign, to obtain $x \geq -1$. Therefore, any real number greater than or equal to -1 satisfies the original inequality.

Coordinate Geometry

Coordinate geometry is the intersection of algebra and geometry. Within this system, the points in a geometric shape are defined using ordered pairs. In the two-dimensional coordinate system, an *x*- and *y*-axis form the *xy-plane*. The *x*-axis is a horizontal scale, and the *y*-axis is a vertical scale. The ordered pair where the axes cross is known as the **origin**. To the right of the origin, the *x* values are positive, and to the left of the origin, the *x* values are negative. The *y* values above the origin are positive, and *y* values below the origin are negative. The axes split the plane into four quadrants, and the first quadrant is where both *x* and *y* values are positive. To plot an ordered pair means to locate the point corresponding to the *x* and *y* coordinates. For example, plotting (4,3) means moving to the right 4 units from 0 in the *x* direction and then moving up 3 units in the *y* direction.

Here is a picture of the *xy*-plane, also known as the **rectangular** or **Cartesian coordinate system**:

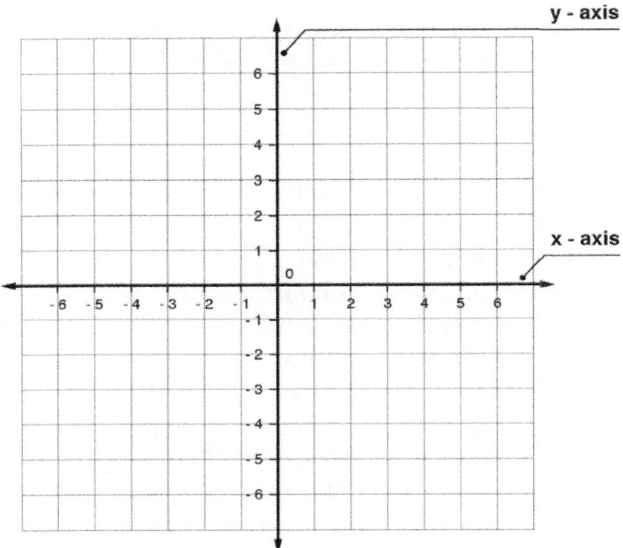

The coordinate system allows us to visualize relationships between equations and geometric figures. For instance, an equation in two variables, *x* and *y*, is represented as a straight line on the *xy* coordinate plane. A solution of an equation in two variables is an ordered pair that satisfies the equation. A graph of an equation can be found by plotting several ordered pairs that are solutions of the equation and then connecting those points with a straight line or smooth curve.

Here is the graph of $4x + y = 8$:

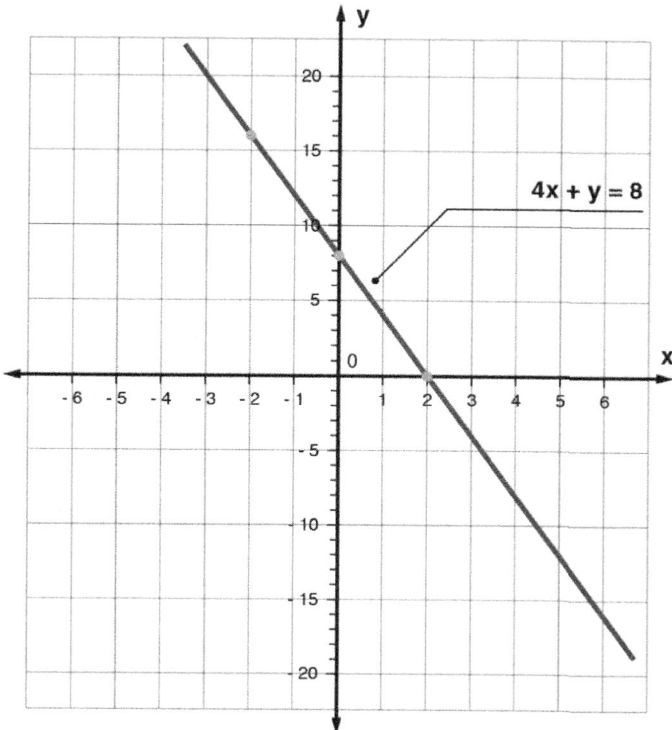

Three ordered pairs that are solutions to the equation were found and plotted. Those points are (–2,16), (0,8), and (2,0). The points were connected using a straight line. Note that the point (0,8) is where the line crosses the *y*-axis. This point is known as the **y-intercept**. The *y*-intercept can always be found by plugging $x = 0$ into the equation. Also, the point (2,0) is where the line crosses the *x*-axis. This point is known as the **x-intercept**, and it can always be found for any equation of a line by plugging $y = 0$ into the equation. The equation above is written in standard form, $Ax + By = C$. Often an equation is written in slope-intercept form, $y = mx + b$, where m represents the slope of the line, and b represents the *y*-intercept. The above equation can be solved for *y* to obtain $y = -4x + 8$, which shows a slope of –4 and a *y*-intercept of 8, meaning the point (0,8).

The **slope** of a line is the measure of steepness of a line, and it compares the vertical change of the line, the **rise**, to the horizontal change of the line, the **run**. The formula for slope of a line through two distinct points (x_1, y_1) and (x_2, y_2) is:

$$m = \frac{y_2 - y_1}{x_2 - x_1}$$

If the line increases from left to right, the slope is positive, and if the line decreases from left to right, as shown above, the slope is negative. If a line is horizontal, like the line representing the equation $y = 5$, the slope is 0. If a line is vertical, like the line representing the equation $x = 2$, the line has **undefined slope**.

In order to graph a function, it can be done the same way as equations. The $f(x)$ represents the dependent variable y in the equation, so replace $f(x)$ with y, and plot some points. For example, the same graph above would be found for the function:

$$f(x) = -4x + 8$$

Graphs other than straight lines also exist. For instance, here are the graphs of $f(x) = x^2$ and $f(x) = x^3$, the squaring and cubic functions.

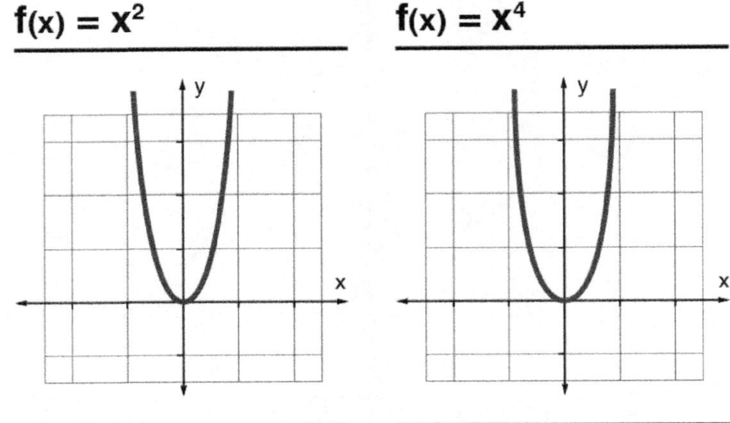

If the equals sign is changed to an **inequality symbol** such as $<. >, \leq$, or $\geq$ in an equation, the result is an inequality. If it is changed to a linear equation in two variables, the result is a linear inequality in two variables. A solution of an inequality in two variables is an ordered pair that satisfies the inequality. For example, (1,3) is a solution of the linear inequality $y \geq x + 1$ because when plugged in, it results in a true statement. The graph of an inequality in two variables consists of all ordered pairs that make the solution true.

A **half-plane** consists of the set of all points on one side of a line in the *xy*-plane, and the solution to a linear inequality is a half-plane. If the inequality consists of $>$ or $<$, the line is dashed, and no solutions actually exist on the line shown. If the inequality consists of $\geq$ or $\leq$, the line is solid, and solutions do exist on the line shown. In order to graph a linear inequality, graph the corresponding equation found by replacing the inequality symbol with an equals sign. Then pick a test point on either side of the line. If that point results in a true statement when plugged into the original inequality, shade in the side containing the test point. If it results in a false statement, shade in the opposite side.

Here is the graph of the inequality $y < x + 1$.

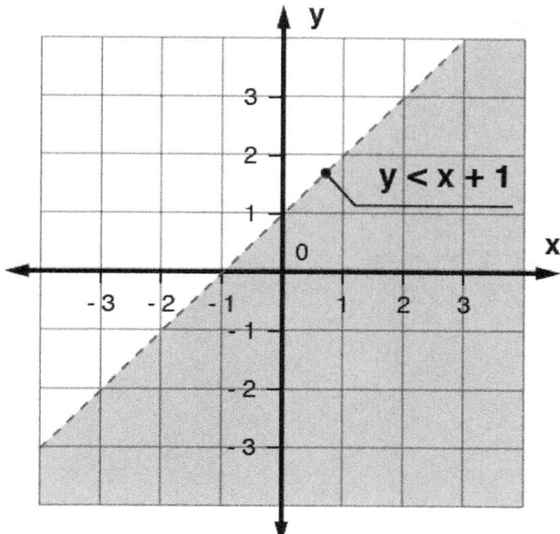

Geometry

Lines

Two lines are **parallel** if they never intersect. Given the equation of two lines, they are parallel if they have the same slope and different *y*-intercepts. If they had the same slope and same *y*-intercept, they would be the same line. Therefore, in order to show two lines are parallel, put them in slope-intercept form, $y = mx + b$, to find *m* and *b*. The two lines:

$$y = 2x + 6 \; and \; 4x - 2y = 6$$

are parallel. The second line in slope intercept is:

$$y = 2x - 3$$

Both lines have the same slope, 2, and different *y*-intercepts.

Two lines are **perpendicular** if they intersect at a right angle. Given the equation of two lines, they are perpendicular if their slopes are negative reciprocals. Therefore, the product of both slopes is equal to – 1. For example, the lines:

$$y = 4x + 1 \; and \; y = -\frac{1}{4}x + 1$$

are perpendicular because their slopes are negative reciprocals. The product of 4 and $-\frac{1}{4}$ is –1.

Circles

A **circle** is defined to be the set of all points the same distance, known as **radius** *r*, from a single point *C*, known as the **center**. A circle measures 360 degrees. The radius is the length from the center to any point on its edge. Multiply the radius times 2, to obtain the **diameter**, which is the distance from any two points on the circle that goes through the center. An **arc** is defined to be all points between any two points on the edge of a circle. A **sector** can be built from an arc and two corresponding radii. Finally, a **central angle** is the angle formed by the intersection of those two radii within a sector. Here is a picture that highlights all of these definitions:

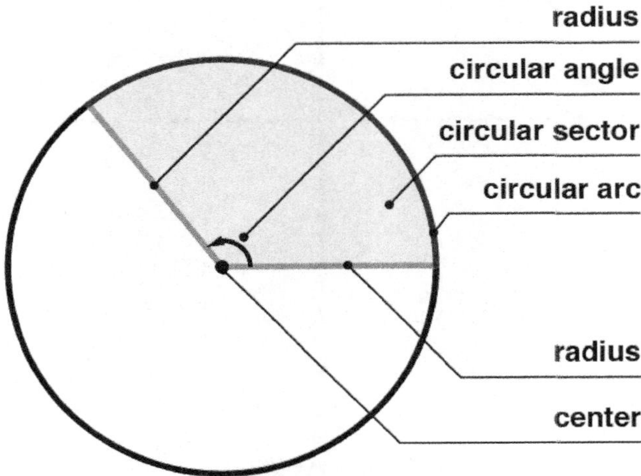

Similar to perimeter, the **circumference** of a circle is equal to the total distance around the outside. The formula for circumference is $C = 2\pi r$, which is the same as $C = \pi d$. The formula for arc length is:

$$2\pi r \frac{\text{central angle of arc meaasurement}}{360}$$

The units for both circumference and arc length are linear units, like inches or centimeters, since they are a measurement of length. The formula for area of a circle is $A = \pi r^2$, and the formula for area of a sector is:

$$\pi r^2 \frac{\text{central angle of arc meaasurement}}{360}$$

Both units of area are square units.

Triangles

Within a triangle, the measurement of all three angles adds up to 180 degrees. There are three types of special triangles. An **equilateral triangle** has three equal sides and three equal angles, which are each 60 degrees. An **isosceles triangle** has two equal sides, and therefore the measurement of the angles opposite the two equal sides are equal as well. Finally, a **right triangle** has one right angle. The side across from the right angle is known as the **hypotenuse**, and the other two sides are known as the **legs**. There is a special type of right triangle known as a 30-60-90 triangle because the three angles in the triangle measure 30, 60, and 90 degrees, respectively.

The legs of this type of triangle have set relationships due to their corresponding angles. Because the sum of the measures of three angles in any triangle is 180 degrees, if only two angles are known inside a triangle, the third can be found by subtracting the sum of the two known quantities from 180. Two angles whose sum is equal to 90 degrees are known as **complementary angles**. For example, angles measuring 72 and 18 degrees are complementary, and each angle is a complement of the other. Finally, two angles whose sum is equal to 180 degrees are known as **supplementary angles**. To find the supplement of an angle, subtract the given angle from 180 degrees. For example, the supplement of an angle that is 50 degrees is:

180 – 50 = 130 degrees

These terms involving angles can be seen in many types of word problems. For example, consider the following problem: The measure of an angle is 60 degrees less than two times the measure of its complement. What is the angle's measure? To solve this, let x be the unknown angle. Therefore, its complement is $90 - x$. The problem gives that:

$$x = 2(90 - x) - 60$$

To solve for x, distribute the 2, and collect like terms. This process results in:

$$x = 120 - 2x$$

Then, use the addition property to add $2x$ to both sides to obtain $3x = 120$. Finally, use the multiplication properties of equality to divide both sides by 3 to get $x = 40$. Therefore, the angle measures 40 degrees. Also, its complement measures 50 degrees.

Quadrilaterals

A **quadrilateral** is any four-sided polygon, such as a square, rectangle, parallelogram, or trapezoid. Basically, a quadrilateral is a closed shape with four sides. A **parallelogram** is a specific type of quadrilateral that has two sets of parallel lines having the same length. A **trapezoid** is a quadrilateral having only one set of parallel sides. A **rectangle** is a parallelogram that has four right angles and two pairs of equal sides, the length and width or the base and height. A **rhombus** is a parallelogram with two acute angles, two obtuse angles, and four equal sides. The acute angles are of equal measure, and the obtuse angles are of equal measure. Finally, a **square** is a rhombus consisting of four right angles with all sides equal in length. It is important to note that some of these shapes share common attributes. For instance, all four-sided shapes are quadrilaterals. All squares are rectangles, but not all rectangles are squares.

Polygons

A **polygon** is any closed figure consisting of three or more line segments. A three-sided polygon is known as a **triangle**, a four-sided polygon is known as a **quadrilateral**, a five-sided polygon is known as a **pentagon**, a six-sided polygon is known as a **hexagon**, an eight-sided polygon is known as an **octagon**, and a ten-sided polygon is known as a **decagon**. In order to calculate the perimeter of any polygon, just add up the length of all of its sides. For a polygon with n sides, the sum of all angles is determined by the following formula:

$$(n - 2) \times 180°$$

For example, a pentagon has five sides, so the sum of all of its angles is:

$$(5 - 2) \times 180° = 540°$$

A **regular polygon** is defined as one in which all sides and angles are of equal measure. For example, a regular three-sided polygon is an equilateral triangle, and a regular four-sided polygon is a square. In a regular polygon, the measure of each angle is found by dividing the sum of all angles by the number of angles. For instance, a regular pentagon contains five angles, and each has measure $\frac{540}{5} = 108$ degrees.

Congruent and Similar Figures

Two figures are **congruent** if they have the same shape and same size, meaning the same angle measurements and equal side lengths. Two figures are **similar** if they have the same angle measurement but not side lengths. In other words, angles are congruent in similar triangles. Therefore, proving figures are congruent involves showing all angles and sides are the same, and proving figures are similar just involves proving the angles are the same.

If two triangles have two corresponding pairs of angles, the triangles are similar because two angles of equal measure implies equality of the third, since all three add up to 180 degrees. The criteria for triangles to be similar also involve proportionality of side lengths. Corresponding sides of two triangles are sides that are in the same location in the two different shapes. In **similar triangles**, corresponding side lengths need to be a constant multiple of each other, meaning the ratios of the lengths of those corresponding lengths are all equal.

This is highlighted in the following example:

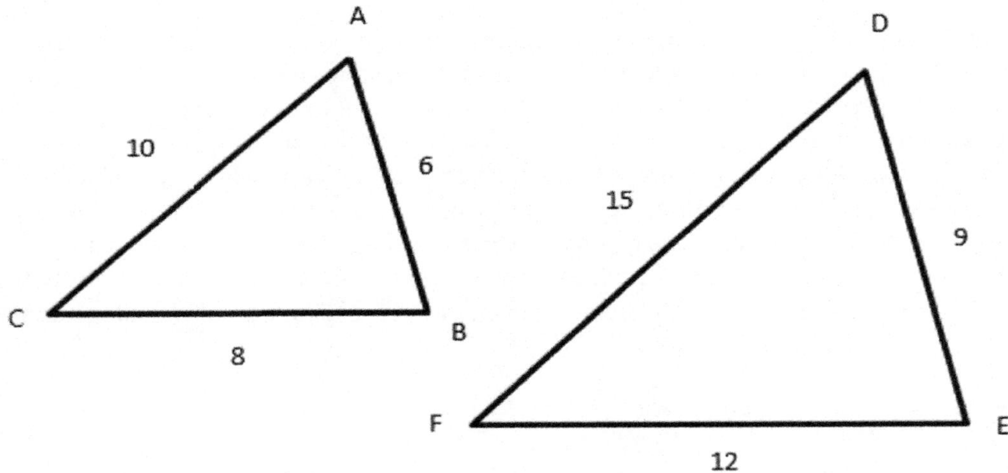

The triangles are similar because each ratio of corresponding sides is equal to 1.5.

Three methods can be used to show two triangles are congruent. First, if the three sides of the first triangle are equal to the sides of the second, the triangles are congruent. Equal sides means equal angles. Second, if two sides and their included angle of the first triangle can be shown to be equal to two sides and their included angle of the second triangle, the triangles are congruent. Third, if two angles and their included side of the first angle can be shown to be equal to two angles and their included side of the second angle, the triangles are congruent.

Relationships in geometric figures other than triangles can be proven using triangle congruence and similarity. If a similar or congruent triangle can be found within another type of geometric figure, their criteria can be used to prove a relationship about a given formula. For example, a rectangle can be broken up into two congruent triangles.

Three-Dimensional Figures

A **rectangular solid** is a six-sided figure with sides that are rectangles. All of the faces meet at right angles, and it looks like a box. Its three measurements are length *l*, width *w*, and height *h*.

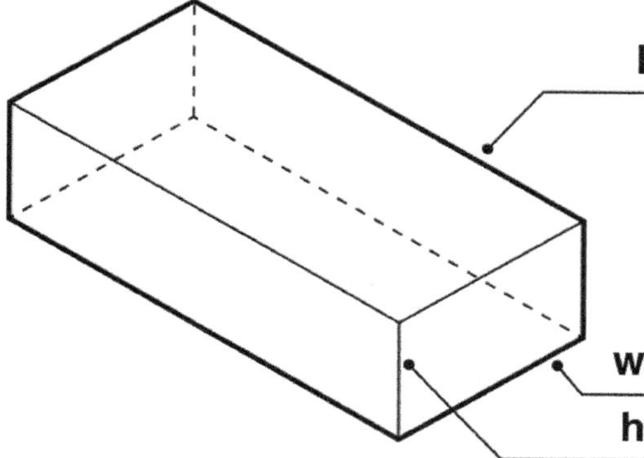

If all sides are equal in a rectangular solid, the solid is known as a **cube**. The cube has six congruent faces that meet at right angles, and each side length is the same and is labeled *s*.

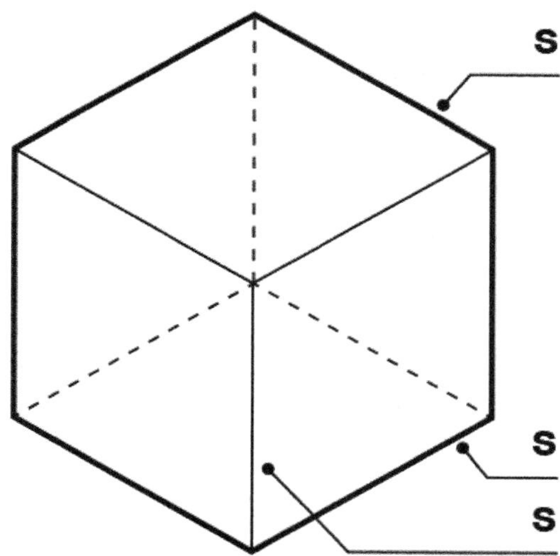

A **cylinder** is a three-dimensional geometric figure consisting of two parallel circles and two parallel lines connecting the ends. The circle has radius *r*, and the cylinder has height *h*.

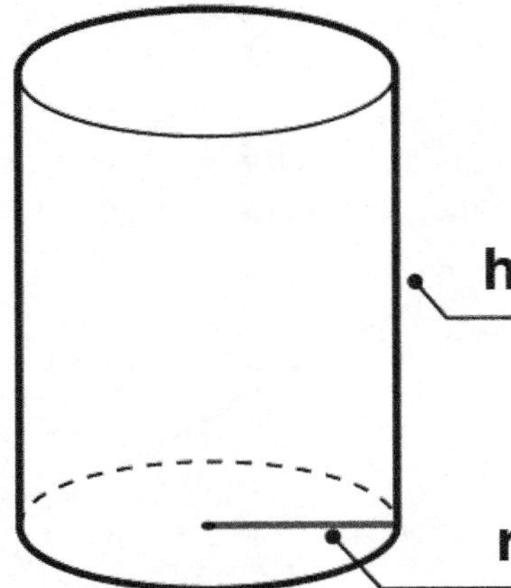

Finally, a **sphere** is a symmetrical three-dimensional shape, where every point on the surface is equal distance from its center. It has a radius *r*.

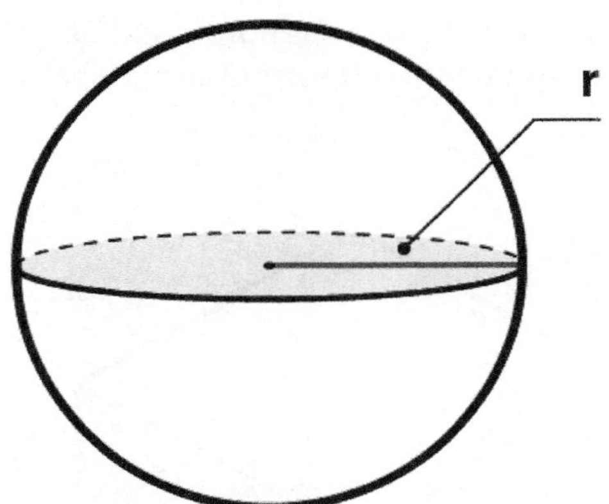

Area

Area is the two-dimensional space covered by an object. These problems may include the area of a rectangle, a yard, or a wall to be painted. Finding the area may be a simple formula, or it may require multiple formulas to be used together. The units for area are square units, such as square meters, square inches, and square miles. Given a square with side length s, the formula for its area is $A = s^2$. Some other formulas for common shapes are shown below.

Shape	Formula	Graphic
Rectangle	$Area = length \times width$	
Triangle	$Area = \frac{1}{2} \times base \times height$	height base
Circle	$Area = \pi \times radius^2$	radius

The following formula, not as widely used as those shown above, but very important, is the area of a trapezoid:

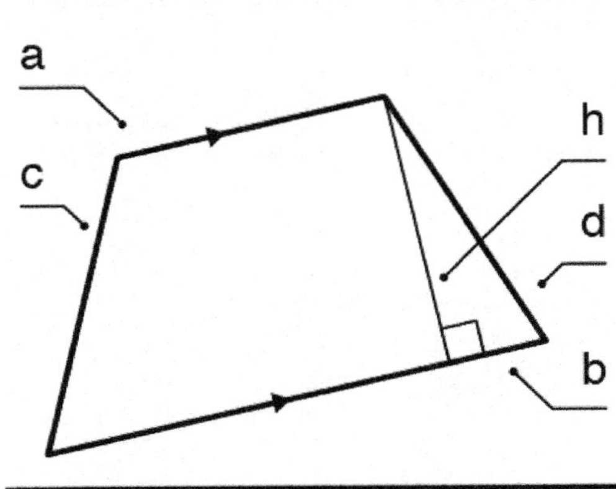

$$A = \frac{1}{2}(a + b)h$$

To find the area of the shapes above, use the given dimensions of the shape in the formula. Complex shapes might require more than one formula. To find the area of the figure below, break the figure into two shapes. The rectangle has dimensions 11 cm by 6 cm. The triangle has dimensions 3 cm by 6 cm. Plug the dimensions into the rectangle formula:

$$A = 11 \times 6$$

Multiplication yields an area of 66 cm^2. The triangle area can be found using the formula:

$$A = \frac{1}{2} \times 4 \times 6$$

Multiplication yields an area of 12 cm^2. Add the areas of the two shapes to find the total area of the figure, which is 78 cm^2.

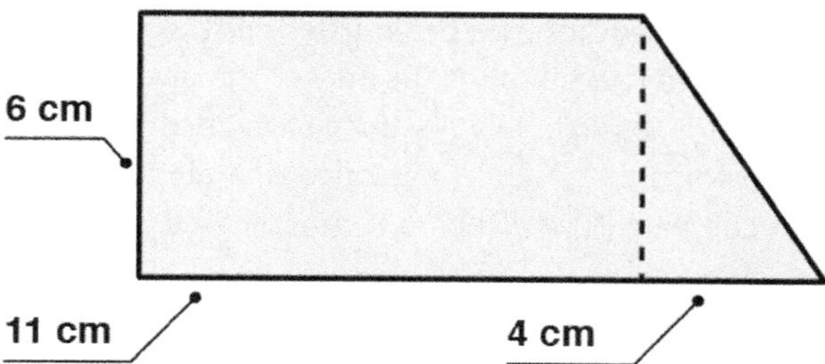

Instead of combining areas, some problems may require subtracting them, or finding the difference.

To find the area of the shaded region in the figure below, determine the area of the whole figure. Then subtract the area of the circle from the whole.

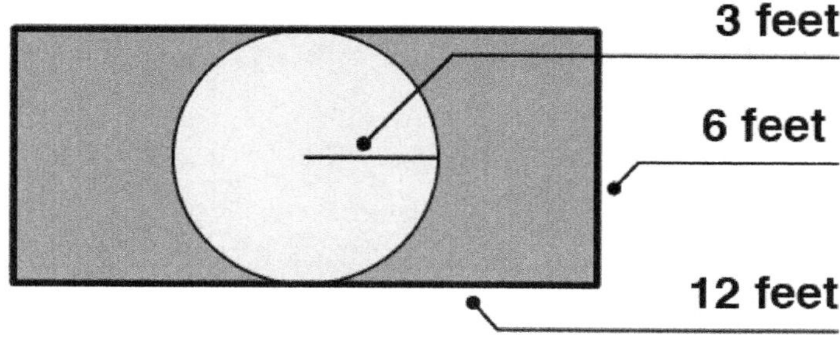

The following formula shows the area of the outside rectangle:

$$A = 12 \times 6 = 72 \text{ ft}^2$$

The area of the inside circle can be found by the following formula:

$$A = \pi(3)^2 = 9\pi = 28.3 \text{ ft}^2$$

As the shaded area is outside the circle, the area for the circle can be subtracted from the area of the rectangle to yield an area of 43.7 ft^2.

Perimeter

Perimeter is the distance around an object. The perimeter of an object can be found by adding the lengths of all sides. Perimeter may be used in problems dealing with lengths around objects such as fences or borders. It may also be used in finding missing lengths or working backwards. If the perimeter is given, but a length is missing, use subtraction to find the missing length.

Given a square with side length s, the formula for perimeter is $P = 4s$. Given a rectangle with length l and width w, the formula for perimeter is:

$$P = 2l + 2w$$

The perimeter of a triangle is found by adding the three side lengths, and the perimeter of a trapezoid is found by adding the four side lengths. The units for perimeter are always the original units of length, such as meters, inches, miles, etc. When discussing a circle, the distance around the object is referred to as its circumference, not perimeter. The formula for circumference of a circle is $C = 2\pi r$, where r represents the radius of the circle. This formula can also be written as $C = d\pi$, where d represents the diameter of the circle.

Volume

Volume is three-dimensional and describes the amount of space that an object occupies, but it's different from area because it has three dimensions instead of two. The units for volume are **cubic units**, such as cubic meters, cubic inches, and cubic millimeters. Volume can be found by using formulas for common objects such as cylinders and boxes.

The following chart shows a diagram and formula for the volume of two objects.

Shape	**Formula**	**Diagram**
Rectangular Prism (box)	$V = length \times width \times height$	length height width
Cylinder	$V = \pi \times radius^2 \times height$	radius height

Volume formulas of these two objects are derived by finding the area of the bottom two-dimensional shape, such as the circle or rectangle, and then multiplying times the height of the three-dimensional shape. Other volume formulas include the volume of a cube with side length s: $V = s^3$; the volume of a sphere with radius r: $V = \frac{4}{3}\pi r^3$; and the volume of a cone with radius r and height h:

$$V = \frac{1}{3}\pi r^2 h$$

If a soda can has a height of 5 inches and a radius on the top of 1.5 inches, the volume can be found using one of the given formulas. A soda can is a cylinder. Knowing the given dimensions, the formula can be completed as follows:

$$V = \pi(radius)^2 \times height$$

$$\pi(1.5 \text{ in})^2 \times 5 \text{ in} = 35.325 \text{ in}^3$$

Notice that the units for volume are inches cubed because it refers to the number of cubic inches required to fill the can.

Pythagorean Theorem

The **Pythagorean theorem** states that given a right triangle, the sum of the squares of the two legs equals the square of the hypotenuse. For example, consider the following right triangle:

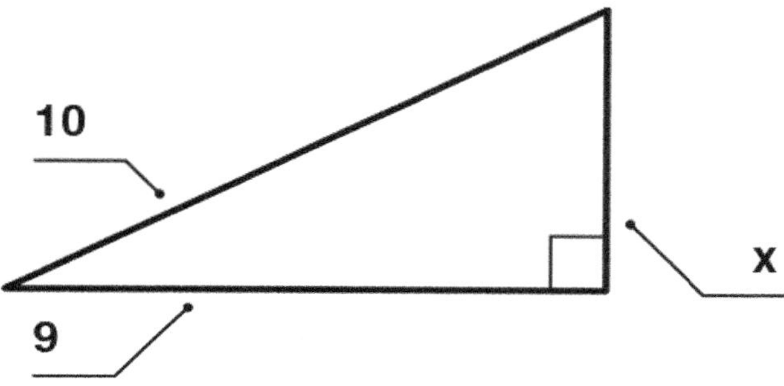

The missing side, x, can be found using the Pythagorean theorem. Since:

$$9^2 + x^2 = 10^2$$

$$81 + x^2 = 100$$

which gives $x^2 = 19$. To solve for x, take the square root of both sides. Therefore, $x = \sqrt{19} = 4.36$, which has been rounded to two decimal places.

Angle Measurement

In geometry, a **line** connects two points and extends indefinitely in both directions beyond each point. If the length is finite, it is known as a **line segment** and has two end points at either end. A **ray** is a straight path that has one end point and extends indefinitely in the other direction. When lines are extended indefinitely, an arrow is used instead of a point. An angle is formed when two rays begin at the same end point and both extend indefinitely in different directions. The common end point is called a **vertex**. **Adjacent angles** are formed from two angles using one shared ray. They are two side-by-side angles that also share an end point.

Angles are measured in degrees, and their measurement assesses rotation. A full rotation equals 360 degrees and represents a circle. Half of a rotation equals 180 degrees and represents a half-circle. Ninety degrees represents a quarter-circle, which is known as a **right angle**. Any angle less than 90 degrees is called an **acute angle**, and any angle greater than 90 degrees is called an **obtuse angle**. Angle measurement is **additive**, meaning if an angle is broken up into two non-overlapping angles, the total measurement of the larger angle is the sum of the two smaller angles.

A **protractor** can be used to measure an angle. Here is a picture of a protractor measuring a right angle:

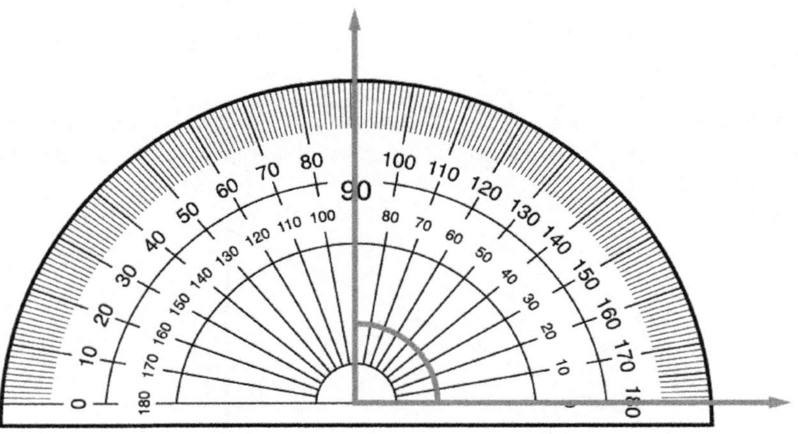

Data Analysis

Descriptive Statistics

Mean, Median, Mode, and Range

One way information can be interpreted from tables, charts, and graphs is through descriptive statistics. The three most common calculations for a set of data are the mean, median, and mode. These three are called measures of central tendency. Measures of central tendency are helpful in comparing two or more different sets of data. The **mean** refers to the average and is found by adding up all values and dividing the total by the number of values. In other words, the mean is equal to the sum of all values divided by the number of data entries. For example, if you bowled a total of 532 points in 4 bowling games, your mean score was $\frac{532}{4} = 133$ points per game. A common application of mean useful to students is calculating what he or she needs to receive on a final exam to receive a desired grade in a class.

The **median** is found by lining up values from least to greatest and choosing the middle value. If there's an even number of values, then the mean of the two middle amounts must be calculated to find the median. For example, the median of the set of dollar amounts $5, $6, $9, $12, and $13 is $9. The median of the set of dollar amounts $1, $5, $6, $8, $9, $10 is $7, which is the mean of $6 and $8. The **mode** is the value that occurs the most. The mode of the data set {1, 3, 1, 5, 5, 8, 10} actually refers to two numbers: 1 and 5. In this case, the data set is bimodal because it has two modes. A data set can have no mode if no amount is repeated. Another useful statistic is range. The **range** for a set of data refers to the difference between the highest and lowest value.

In some cases, some numbers in a list of data might have weights attached to them. In that case, a **weighted mean** can be calculated. A common application of a weighted mean is GPA. In a semester, each class is assigned a number of credit hours, its weight, and at the end of the semester each student receives a grade. To compute GPA, an A is a 4, a B is a 3, a C is a 2, a D is a 1, and an F is a 0. Consider a student that takes a 4-hour English class, a 3-hour math class, and a 4-hour history class and receives all B's. The weighted mean, GPA, is found by multiplying each grade times its weight, number of credit hours, and dividing by the total number of credit hours. Therefore, the student's GPA is:

$$\frac{3 \times 4 + 3 \times 3 + 3 \times 4}{11} = \frac{33}{1} = 3.0.$$

Standard Deviation, Quartiles, and Percentiles

A set of data can be described using its **standard deviation**, or spread. It measures how spread apart the data is within the set. The standard deviation actually quantifies the amount of variation with respect to the mean of the dataset. A lower standard deviation shows that the dataset does not differ much from the mean. A standard deviation equal to 0 means that every value is the same in a dataset. Therefore, a larger standard deviation shows that the dataset, as a whole, varies largely from the mean. The formula for sample standard deviation of a sample dataset is:

$$s = \sqrt{\frac{\Sigma(x - \bar{x})^2}{n - 1}}$$

where x is each value in the dataset, $\bar{x}$ is the mean, and n is the total number of data points in the set.

A dataset can be broken up into four equal parts. The three **quartiles** Q_1, Q_2, and Q_3 split up the data into four equal parts. Q_1 is the first quartile, and one-quarter of the data falls on or below it. Q_2 is the second quartile, also the median, and one-half of the data falls on or below it. Q_3 is the third quartile, and three-quarters of the data falls on or below it. The **interquartile range (IQR)** of a dataset gives the range of the middle 50 percent of the data, and its formula is:

$$IQR = Q_3 - Q_1$$

Similar to quartiles, **deciles** divide a dataset into ten equal parts, and **percentiles** divide a dataset into one hundred equal parts. For example, the 90th percentile refers to splitting up a dataset into the bottom 90 percent of the data and the top 10 percent of the data.

Interpretation of Data

Tables, **charts**, and **graphs** can be used to convey information about different variables. They are all used to organize, categorize, and compare data, and they all come in different shapes and sizes. Each type has its own way of showing information, whether it is in a column, shape, or picture. To answer a question relating to a table, chart, or graph, some steps should be followed. First, the problem should be read thoroughly to determine what is being asked to determine what quantity is unknown. Then, the title of the table, chart, or graph should be read. The title should clarify what data is actually being summarized in the table. Next, look at the key and labels for both the horizontal and vertical axes, if they are given. These items will provide information about how the data is organized. Finally, look to see if there is any more labeling inside the table. Taking the time to get a good idea of what the table is summarizing will be helpful as it is used to interpret information.

Tables are a good way of showing a lot of information in a small space. The information in a table is organized in columns and rows. For example, a table may be used to show the number of votes each candidate received in an election. By interpreting the table, one may observe which candidate won the election and which candidates came in second and third. In using a bar chart to display monthly rainfall amounts in different countries, rainfall can be compared between countries at different times of the year. **Graphs** are also a useful way to show change in variables over time, as in a line graph, or percentages of a whole, as in a pie graph.

The table below relates the number of items to the total cost. The table shows that 1 item costs $5. By looking at the table further, 5 items cost $25, 10 items cost $50, and 50 items cost $250. This cost can be extended for any number of items. Since 1 item costs $5, then 2 items would cost $10. Though this information isn't in the table, the given price can be used to calculate unknown information.

Number of Items	1	5	10	50
Cost ($)	5	25	50	250

A bar graph is a graph that summarizes data using bars of different heights. It is useful when comparing two or more items or when seeing how a quantity changes over time. It has both a horizontal and vertical axis. Interpreting **bar graphs** includes recognizing what each bar represents and connecting that to the two variables. The bar graph below shows the scores for 6 people on 3 different games. The color of the bar shows which game each person played, and the height of the bar indicates their score for that game. Emily scored 42 on game 3, and Olivia scored 38 on game 3. By comparing the bars, it's obvious that

Olivia scored lower than Emily. By observing the exact scores, Olivia scored 4 points lower on game 3 than Emily.

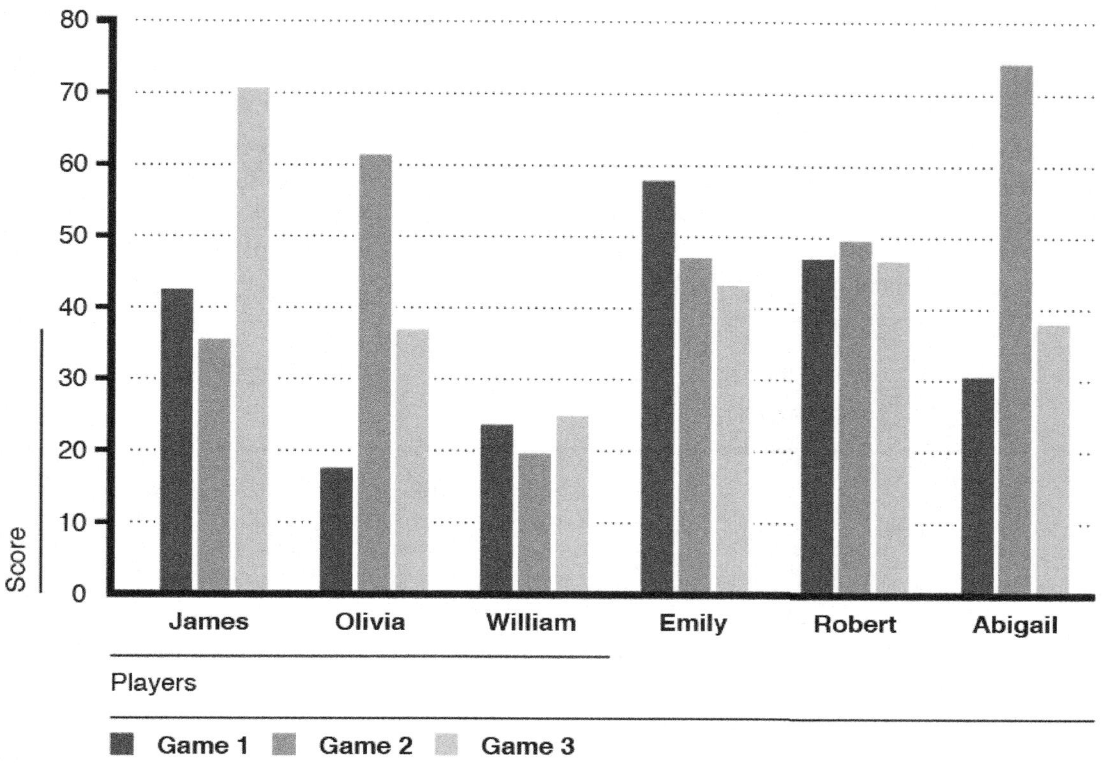

A line graph is a way to compare two variables. Each variable is plotted along an axis, and the graph contains both a horizontal and a vertical axis. On a **line graph**, the line indicates a continuous change. The change can be seen in how the line rises or falls, known as its slope, or rate of change. Often, in line graphs, the horizontal axis represents a variable of time. Audiences can quickly see if an amount has grown or decreased over time. The bottom of the graph, or the x-axis, shows the units for time, such as days, hours, months, etc. If there are multiple lines, a comparison can be made between what the two lines represent.

For example, the following line graph shows the change in temperature over five days. The lighter line represents the high, and the darker line represents the low for each day. Looking at the lighter line alone, the high decreases from Monday to Tuesday, then increases from Tuesday to Wednesday. The low temperatures have a similar trend, shown by the darker line. The range in temperatures each day can also be calculated by finding the difference between the lighter line and the darker line on a particular day. On Wednesday, the range was 13 degrees, from 62 to 75°F.

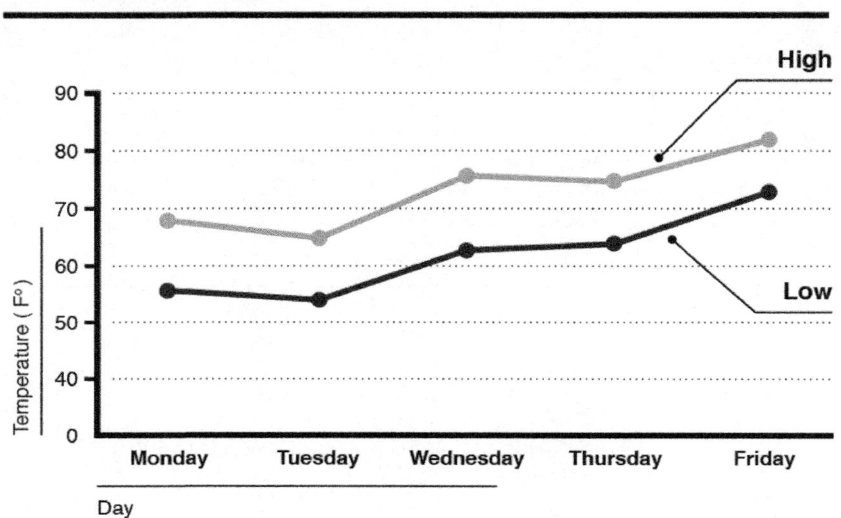

Pie charts are used to show percentages of a whole, as each category is given a piece of the pie, and together all the pieces make up a whole. They are a circular representation of data which are used to highlight numerical proportion. It is true that the arc length of each pie slice is proportional to the amount it individually represents. When a pie chart is shown, an audience can quickly make comparisons by comparing the sizes of the pieces of the pie. They can be useful for comparison between different categories. The following pie chart is a simple example of three different categories shown in comparison to each other.

The labels show that black represents dogs, dark gray represents other pets, and light gray represents cats. As the pie is cut into three equal pieces, each value represents just more than 33 percent, or $\frac{1}{3}$ of the whole. Dogs and cats may be combined to represent $\frac{2}{3}$ of the whole.

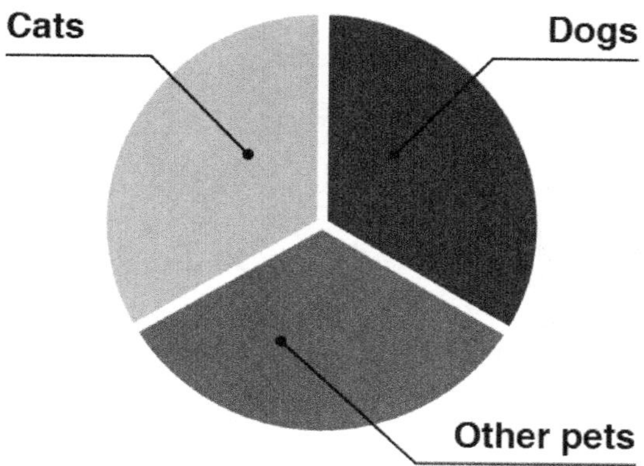

Since circles have 360 degrees, they are used to create pie charts. Because each piece of the pie is a percentage of a whole, that percentage is multiplied times 360 to get the number of degrees each piece represents. In the example above, each piece is $\frac{1}{3}$ of the whole, so each piece is equivalent to 120 degrees. Together, all three pieces add up to 360 degrees.

Stacked bar graphs, also used fairly frequently, are used when comparing multiple variables at one time. They combine some elements of both pie charts and bar graphs, using the organization of bar graphs and the proportionality aspect of pie charts. The following is an example of a stacked bar graph that represents the number of students in a band playing drums, flutes, trombones, and clarinet. Each bar graph is broken up further into girls and boys.

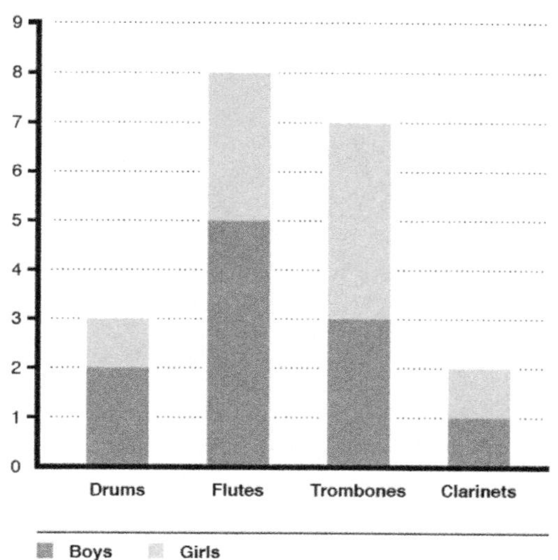

To determine how many boys play trombone, refer to the dark gray portion of the trombones bar, resulting in 3 students.

A **scatterplot** is another way to represent paired data. It uses Cartesian coordinates, like a line graph, meaning it has both a horizontal and vertical axis. Each data point is represented as a dot on the graph. The dots are never connected with a line. For example, the following is a scatterplot showing people's age versus height.

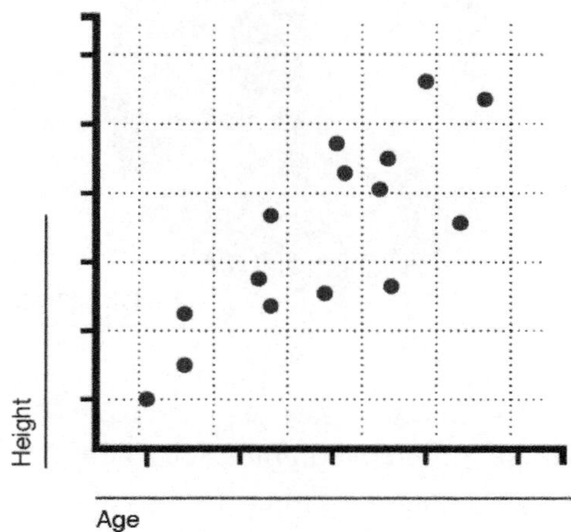

A scatterplot, also known as a **scattergram,** can be used to predict another value and to see if an association, known as a **correlation,** exists between a set of data. If the data resembles a straight line, the data is **associated**. The following is an example of a scatterplot in which the data does not seem to have an association:

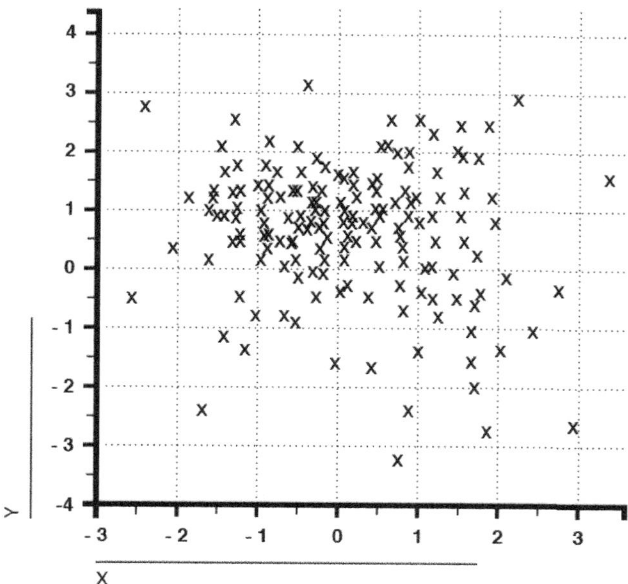

Elementary Probability

A **probability experiment** is an action that causes specific results, such as counts or measurements. The result of such an experiment is known as an **outcome**, and the set of all potential outcomes is known as the **sample space**. An **event** consists of one or more of those outcomes. For example, consider the probability experiment of tossing a coin and rolling a six-sided die. The coin has two possible outcomes—a heads or a tails—and the die has six possible outcomes—rolling each number 1–6. Therefore, the sample space has twelve possible outcomes: a heads or a tails paired with each roll of the die.

A **simple event** is an event that consists of a single outcome. For instance, selecting a queen of hearts from a standard fifty-two-card deck is a simple event; however, selecting a queen is not a simple event because there are four possibilities.

Classical, or **theoretical, probability** is when each outcome in a sample space has the same chance to occur. The probability for an event is equal to the number of outcomes in that event divided by the total number of outcomes in the sample space. For example, consider rolling a six-sided die. The probability of rolling a 2 is $\frac{1}{6}$, and the probability of rolling an even number is $\frac{3}{6}$, or $\frac{1}{2}$, because there are three even numbers on the die. This type of probability is based on what should happen in theory but not what actually happens in real life.

Empirical probability is based on actual experiments or observations. For instance, if a die is rolled eight times, and a 1 is rolled two times, the empirical probability of rolling a 1 is $\frac{2}{8} = \frac{1}{4}$, which is higher than the theoretical probability. The Law of Large Numbers states that as an experiment is completed repeatedly, the empirical probability of an event should get closer to the theoretical probability of an event.

Probabilities range from 0 to 1. The closer the probability of an event occurring is to 0, the less likely it will occur. The closer it is to 1, the more likely it is to occur.

The **addition rule** is necessary to find the probability of event *A* or event *B* occurring or both occurring at the same time. If events *A* and *B* are **mutually exclusive** or **disjoint**, which means they cannot occur at the same time:

$$P(A \text{ or } B) = P(A) + P(B)$$

If events *A* and *B* are not mutually exclusive, $P(A \text{ or } B) = P(A) + P(B) - P(A \text{ and } B)$ where $P(A \text{ and } B)$ represents the probability of event *A* and *B* both occurring at the same time. An example of two events that are mutually exclusive are rolling a 6 on a die and rolling an odd number on a die. The probability of rolling a 6 or rolling an odd number is:

$$\frac{1}{6} + \frac{3}{6} = \frac{4}{6} = \frac{2}{3}$$

Rolling a 6 and rolling an even number are not mutually exclusive because there is some overlap. The probability of rolling a 6 or rolling an even number is:

$$\frac{1}{6} + \frac{3}{6} - \frac{1}{6} = \frac{3}{6} = \frac{1}{2}.$$

Conditional Probability

The **multiplication rule** is necessary when finding the probability that an event *A* occurs in a first trial and event *B* occurs in a second trial, which is written as $P(A \text{ and } B)$. This rule differs if the events are independent or dependent. Two events *A* and *B* are **independent** if the occurrence of one event does not affect the probability that the other will occur. If *A* and *B* are not independent, they are **dependent**, and the outcome of the first event somehow affects the outcome of the second. If events *A* and B are independent, $P(A \text{ and } B) = P(A)P(B)$, and if events *A* and *B* are dependent, $P(A \text{ and } B) = P(A)P(B|A)$, where $P(B|A)$ represents the probability event *B* occurs given that event *A* has already occurred.

$P(B|A)$ represents **conditional probability**, or the probability of event *B* occurring given that event *A* has already occurred. $P(B|A)$ can be found by dividing the probability of events *A* and *B* both occurring by the probability of event *A* occurring using the formula,

$$P(B|A) = \frac{P\ (A \text{ and } B)}{P(A)}$$

and represents the total number of outcomes remaining for *B* to occur after *A* occurs. This formula is derived from the multiplication rule with dependent events by dividing both sides by $P(A)$. Note that $P(B|A)$ and $P(A|B)$ are not the same. The first quantity shows that event *B* has occurred after event *A*, and the second quantity shows that event *A* has occurred after event *B*. To incorrectly interchange these ideas is known as **confusion of the inverse**.

Consider the case of drawing two cards from a deck of fifty-two cards. The probability of pulling two queens would vary based on whether the initial card was placed back in the deck for the second pull. If the card is placed back in, the probability of pulling two queens is:

$$\frac{4}{52} \times \frac{4}{52} = 0.00592$$

If the card is not placed back in, the probability of pulling two queens is:

$$\frac{4}{52} \times \frac{3}{51} = 0.00452$$

When the card is not placed back in, both the numerator and denominator of the second probability decrease by 1. This is due to the fact that, theoretically, there is one less queen in the deck, and there is one less total card in the deck as well.

Conditional probability is used frequently when probabilities are calculated from tables. Two-way frequency tables display data with two variables and highlight the relationships between those two variables. They are often used to summarize survey results and are also known as **contingency tables**. Each cell shows a count pertaining to that individual variable pairing, known as a **joint frequency**, and the totals of each row and column also are in the tables. Consider the following two-way frequency table:

	70 or older	**69 or younger**	**Totals**
Women	20	40	60
Men	5	35	40
Total	25	75	100

This table shows the breakdown of ages and sexes of 100 people in a particular village. Consider a randomly selected villager. The probability of selecting a male 69 years old or younger is $\frac{35}{100}$ because there are 35 males under the age of 70 and 100 total villagers.

Probability Distributions

A **discrete random variable** is a set of values that is either finite or countably infinite. If there are infinitely many values, being **countable** means that each individual value can be paired with a natural number. For example, the number of coin tosses before getting heads could potentially be infinite, but the total number of tosses is countable. Each toss refers to a number, like the first toss, second toss, etc. A **continuous random variable** has infinitely many values that are not countable. The individual items cannot be enumerated; an example of such a set is any type of measurement. There are infinitely many heights of human beings due to decimals that exist within each inch, centimeter, millimeter, etc.

Each type of variable has its own **probability distribution**, which calculates the probability for each potential value of the random variable. Probability distributions exist in tables, formulas, or graphs. The expected value of a random variable represents what the mean value should be in either a large sample size or after many trials. According to the **Law of Large Numbers**, after many trials, the actual mean and that of the probability distribution should be very close to the expected value. The **expected value** is a weighted average that is calculated as $E(X) = \sum x_i p_i$, where x_i represents the value of each outcome, and p_i represents the probability of each outcome. The expected value if all of the probabilities are equal is:

$$E(X) = \frac{x_1 + x_2 + \cdots + x_n}{n}$$

Expected value is often called the **mean of the random variable** and is known as a **measure of central tendency** like mean and mode.

A **binomial probability distribution** is a probability distribution that adheres to some important criteria. The distribution must consist of a fixed number of trials where all trials are independent, each trial has an outcome classified as either a success or a failure, and the probability of a success is the same in each trial. Within any binomial experiment, *x* is the number of resulting successes, *n* is the number of trials, *P* is the probability of success within each trial, and $Q = 1 - P$ is the probability of failure within each trial. The probability of obtaining *x* successes within *n* trials is:

$$\binom{n}{x} P^x (1 - P)^{n-x}$$

where

$$\binom{n}{x} = \frac{n!}{x!\,(n - x)!}$$

is called the **binomial coefficient**. A binomial probability distribution could be used to find the probability of obtaining exactly two heads on five tosses of a coin.

In the formula, *x* = 2, *n* = 5, *P* = 0.5, and *Q* = 0.5.

A **uniform probability distribution** exists when there is constant probability. Each random variable has equal probability, and its graph is a rectangle because the height, representing the probability, is constant.

Finally, a **normal probability distribution** has a graph that is symmetric and bell-shaped; an example using body weight is shown here:

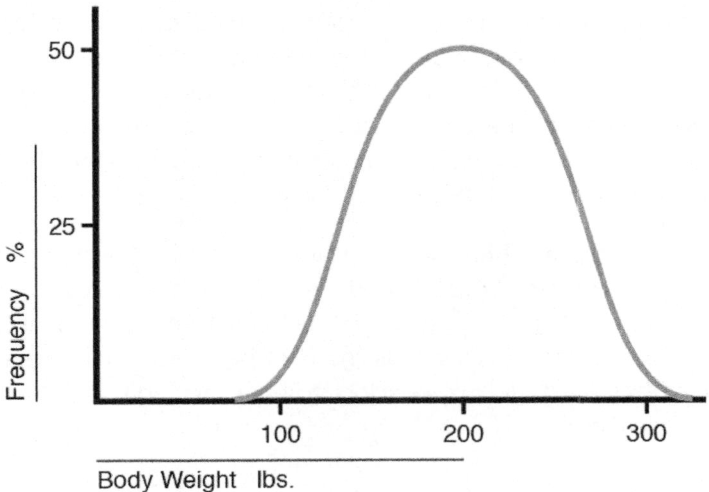

Population percentages can be estimated using normal distributions. For example, the probability that a data point will be less than the mean is 50 percent. The Empirical Rule states that 68 percent of the data falls within 1 standard deviation of the mean, 95 percent falls within 2 standard deviations of the mean, and 99.7 percent falls within 3 standard deviations of the mean. A **standard normal distribution** is a normal distribution with a mean equal to 0 and standard deviation equal to 1. The area under the entire curve of a standard normal distribution is equal to 1.

Counting Methods

The total number of events in the sample space must be known to solve probability problems. Different methods can be used to count the number of possible outcomes, depending on whether different arrangements of the same items are counted only once or separately. **Permutations** are arrangements in which different sequences are counted separately. Therefore, order matters in permutations. **Combinations** are arrangements in which different sequences are not counted separately. Therefore, order does not matter in combinations. For example, if 123 is considered different from 321, permutations would be discussed. However, if 123 is considered the same as 321, combinations would be considered.

If the sample space contains *n* different permutations of *n* different items and all of them must be selected, there are $n!$ different possibilities. For example, five different books can be rearranged 5! = 120 times. The probability of one person randomly ordering those five books in the same way as another person is $\frac{1}{120}$. A different calculation is necessary if a number less than *n* is to be selected or if order does not matter. In general, the notation $P(n,r)$ represents the number of ways to arrange *r* objects from a set of *n* if order does matter, and:

$$P(n,r) = \frac{n!}{(n-r)!}$$

Therefore, in order to calculate the number of ways five books can be arranged in three slots if order matters, plug *n* = 5 and *r* = 3 in the formula to obtain:

$$P(5,3) = \frac{5!}{(5-3)!} = \frac{5!}{2!} = 60$$

Secondly, $C(n,r)$ represents the total number of *r* combinations selected out of *n* items when order does not matter, and:

$$C(n,r) = \frac{n!}{(n-r)!\ r!}$$

Therefore, the number of ways five books can be arranged in three slots if order does not matter is:

$$C(5,3) = \frac{5!}{(5-3)!\ 3!} = 10$$

The following relationship exists between permutations and combinations:

$$C(n,r) = \frac{P(n,r)}{r!}$$

Sets of numbers and other similarly organized data can also be represented graphically. Venn diagrams are a common way to do so. A **Venn diagram** represents each set of data as a circle. The circles overlap, showing that each set of data is overlapping. A Venn diagram is also known as a **logic diagram** because it visualizes all possible logical combinations between two sets. Common elements of two sets are represented by the area of overlap. The following is an example of a Venn diagram of two sets A and B:

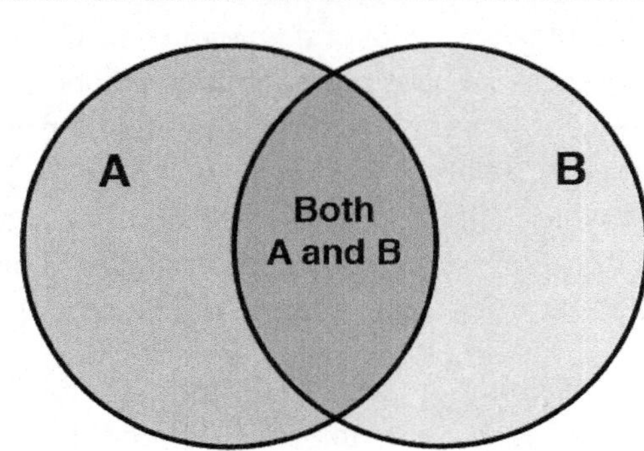

Another name for the area of overlap is the **intersection**. The intersection of A and B, $A \cap B$, contains all elements that are in both sets A and B. The union of A and B, $A \cup B$, contains all elements that are in either set A or set B. Finally, the complement of $A \cup B$ is equal to all elements that are not in either set A or set B. These elements are placed outside of the circles.

The following is an example of a Venn diagram in which 24 students were surveyed asking if they had brothers or sisters or both. Ten students only had brothers, 7 students only had sisters, and 5 had both brothers and sisters. This number 5 is the intersection and is placed where the circles overlap. Two students did not have a cat or a dog. Two is therefore the complement and is placed outside of the circles.

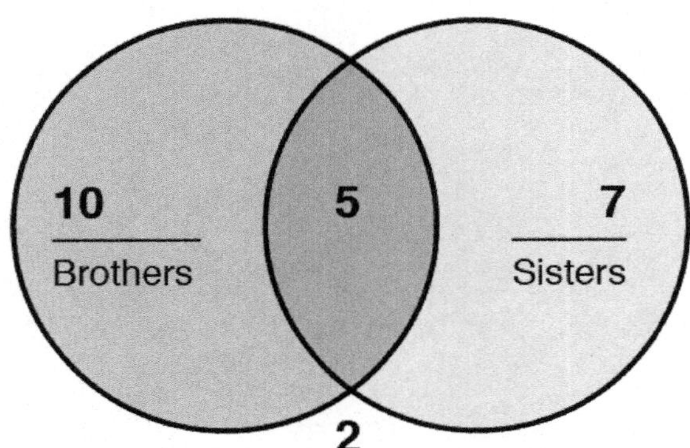

Venn diagrams can have more than two sets of data. The more circles, the more logical combinations are represented by the overlapping. The following is a Venn diagram that represents students who like the colors green, pink, or blue. There were 30 students surveyed. The innermost region represents those students that like green, pink, and blue. Therefore, 2 students like all three. In this example, all students like at least one of the colors, so no one exists in the complement.

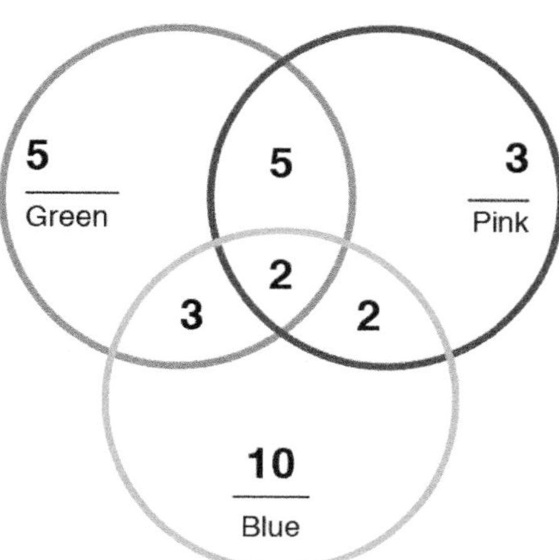

Venn diagrams are typically not drawn to scale, but if they are and their area is proportional to the amount of data it represents, it is known as an **area-proportional** Venn diagram.

Practice Questions

For each of questions 1-5, compare Quantity A to Quantity B, using additional information presented in or above the two quantities. Select one answer choice for each question.

1.

Quantity A	Quantity B
The result of dividing 24 by $\frac{8}{5}$	The product of $\frac{14}{15}$ and $\frac{2}{5}$

a. Quantity A is greater
b. Quantity B is greater
c. The two quantities are equal
d. The relationship cannot be determined from the information given.

2. Use the graph below to compare the two quantities.

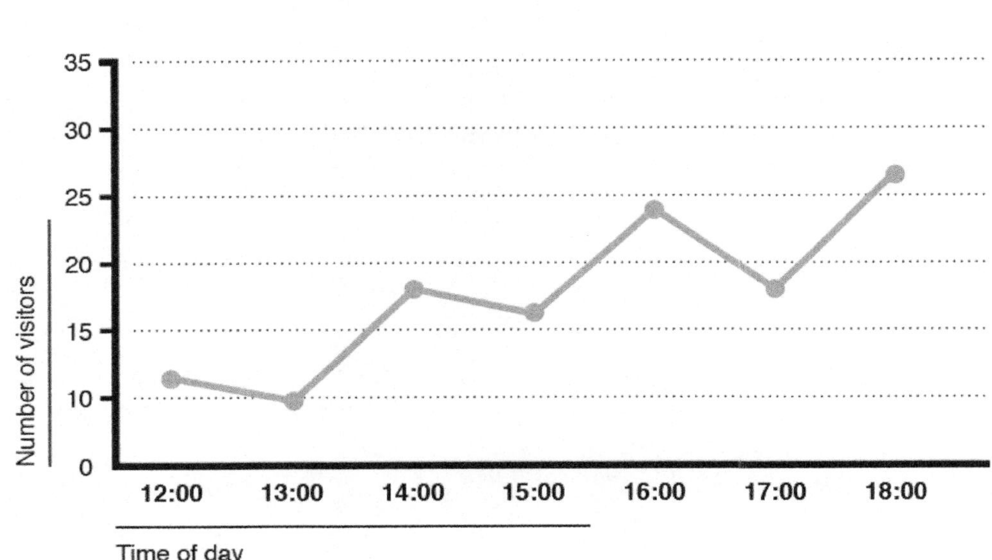

Quantity A	Quantity B
The mean number of visitors for the first four hours	The median of the number of visitors for the total seven hours

a. Quantity A is greater
b. Quantity B is greater
c. The two quantities are equal
d. The relationship cannot be determined from the information given.

3.

Quantity A

The value of $\frac{12}{60}$ converted to a percentage

Quantity B

The percentage of students who live on campus if 40 students out of 200 live on campus.

a. Quantity A is greater
b. Quantity B is greater
c. The two quantities are equal
d. The relationship cannot be determined from the information given.

4.

Quantity A

$8^2 - 3(4 + 5 \times 2)$

Quantity B

$(\sqrt{81} - \sqrt{49}) \times (72 \div 9)$

a. Quantity A is greater
b. Quantity B is greater
c. The two quantities are equal
d. The relationship cannot be determined from the information given.

5. $y = 3x^2 - 9x + 6$

Quantity A

The value of x

Quantity B

The value of y

a. Quantity A is greater
b. Quantity B is greater
c. The two quantities are equal
d. The relationship cannot be determined from the information given.

Questions 6-20 have several different formats. Unless otherwise directed, select a single answer choice for each question. For Numeric Entry questions, follow the instructions below.

Numeric Entry Questions

Enter your answer in the box(es) below the question.

- Your answer may be an integer, a decimal, or a fraction, and it may be negative.
- If a question asks for a fraction, there will be two boxes. One is for the numerator and one is for the denominator.
- Equivalent forms of the value, such as 1.5 and 1.50, are all correct. Fractions do not need to be reduced to lowest terms.
- Enter the exact answer unless your question asks you to round your answer.

6. What is the correct sum of $\frac{14}{15}$ and $\frac{2}{5}$?

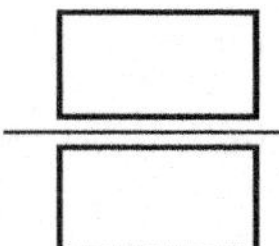

7. Gina took an algebra test last Friday. There were 35 questions, and she answered 60% of them correctly. How many correct answers did she have?

8. A card is drawn from a standard deck of 52 cards. What is the probability that the card is a king or queen?

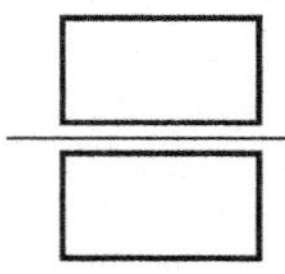

9. A grocery store sold 48 bags of apples in one day. If 9 of the bags contained Granny Smith apples and the rest contained Red Delicious apples, what is the ratio of bags of Granny Smith to bags of Red Delicious that were sold?

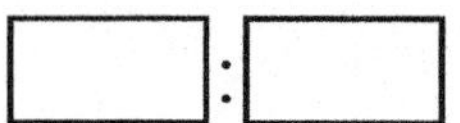

10. What is the area of the darker shaded region?

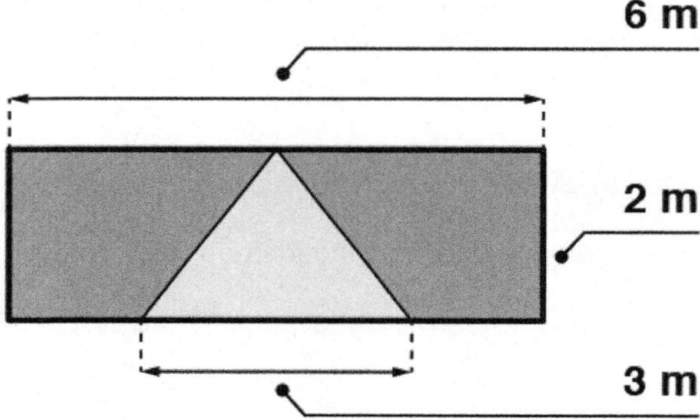

m^2

11. What is the solution to the equation $3(x + 2) = 14x - 5$?

a. $x = 1$
b. No solution
c. $x = 0$
d. All real numbers

12. Two consecutive integers exist such that the sum of three times the first and two less than the second is equal to 411. What are those integers?

a. 102 and 103
b. 104 and 105
c. 103 and 104
d. 100 and 101

13. A car manufacturer usually makes 15,412 SUVs, 25,815 station wagons, 50,412 sedans, 8,123 trucks, and 18,312 hybrids a month. About how many cars are manufactured each month?

a. 120,000
b. 200,000
c. 300,000
d. 12,000

14. Erin and Katie work at the same ice cream shop. Together, they always work less than 21 hours a week. In a week, if Katie worked two times as many hours as Erin, how many hours could Erin work?

a. Less than 7 hours
b. Less than or equal to 7 hours
c. More than 7 hours
d. Less than 8 hours

15. What is the volume of the cylinder below?

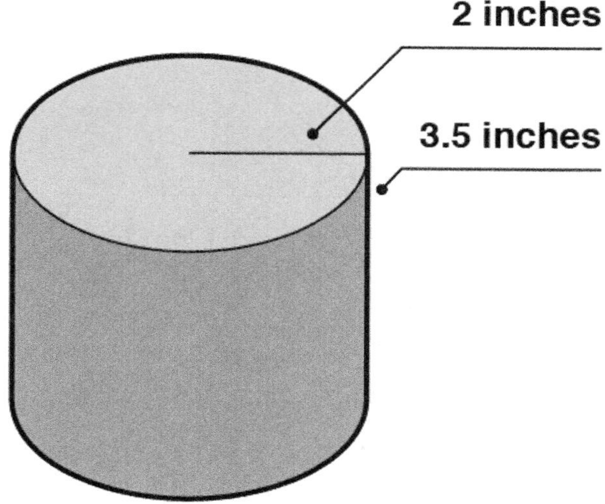

a. 18.84 in^3
b. 45.00 in^3
c. 70.43 in^3
d. 43.96 in^3

16. From the chart below, which genres are preferred by more men than women? Select all that apply.

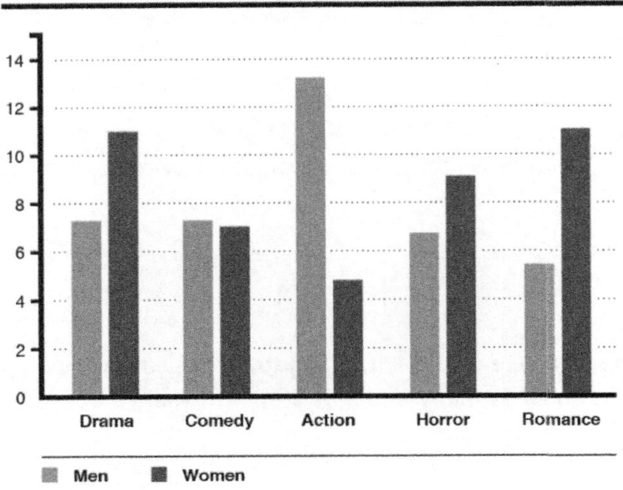

a. Comedy
b. Drama
c. Action
d. Romance
e. Horror

17. Which of the following is the mode for the grades shown in the chart below? Select all that apply.

Science Grades	
Jerry	65
Bill	97
Anna	80
Beth	95
Sara	85
Ben	65
Sally	80
George	98
Crystal	92
Jordan	98

a. 65
b. 89
c. 86
d. 90
e. 80
f. 92
g. 98

18. Which of the following are factors of 54? Select all that apply.
   a. 6
   b. 7
   c. 12
   d. 18
   e. 2
   f. 4
   g. 54

19. Which of the following are true of the following graph? Select all that apply.

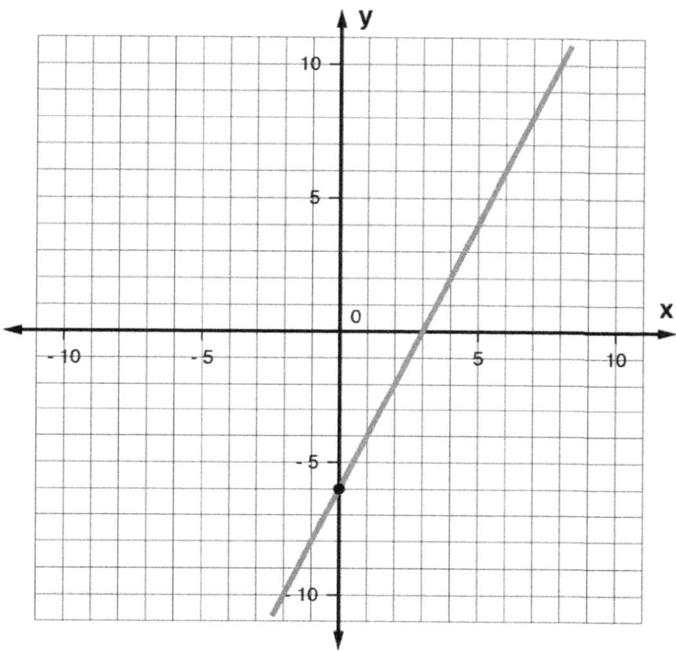

   a. The slope of the line is −2.
   b. The y-intercept is 3.
   c. The slope of the line is 2.
   d. The equation for the line is $y = -2x + 3$.
   e. The y-intercept is −6.
   f. The equation for the line is $y = 2x - 6$.

20. In the given sequence $a_1, a_2, a_3, a_4, \ldots, a_n$, where $a_1 = 4$, n is a positive integer, and $a_{n+1} = 2a_n + 3$, which of the following could be a value of a_n? Select all that apply.
   a. 53
   b. 20
   c. 11
   d. 14
   e. 25
   f. 42

Answer Explanations

1. A: Quantity A is greater because division is completed by multiplying times the reciprocal. Therefore:

$$24 \div \frac{8}{5} = \frac{24}{1} \times \frac{5}{8}$$

$$\frac{3 \times 8}{1} \times \frac{5}{8} = \frac{15}{1} = 15$$

Quantity B is:

$$\frac{14}{15} \times \frac{2}{5} = \frac{28}{75}$$

2. B: Quantity B is greater. The median of the total seven hours can be found by arranging the numbers in order (10, 12, 16, 18, 18, 24, 26) and finding the number in the middle which is 18. The mean for the number of visitors during the first 4 hours is 14. The mean is found by calculating the average for the four hours. Adding up the total number of visitors during those hours gives:

$$12 + 10 + 18 + 16 = 56$$

Dividing total visitors by four hours gives average visitors per hour:

$$56 \div 4 = 14$$

3. C: The two quantities are equal. The fraction $\frac{12}{60}$ can be reduced to $\frac{1}{5}$, in lowest terms. Then, it must be converted to a decimal. Dividing 1 by 5 results in 0.2.

Then, to convert to a percentage, move the decimal point two units to the right and add the percentage symbol. The result is 20%. If 40 students out of 200 live on campus, then the corresponding fraction to represent this situation is $\frac{40}{200}$. This reduces to $\frac{1}{5}$ and can be converted to a percentage of 20%.

4. A: Quantity A is greater. Using order of operations, the expression for Quantity A can be solved as follows:

$$8^2 - 3(4 + 5 \times 2)$$

$$64 - 3(14) = 64 - 42 = 22$$

Using order of operations, the expression for Quantity B can be solved as follows:

$$(\sqrt{81} - \sqrt{49}) \times (72 \div 9)$$

$$(9 - 7) \times (8)$$

$$2 \times 8 = 16$$

5. D: The relationship cannot be determined. This can be determined by substituting numbers for x. If you substitute 1 for x, the value of y is:

$$y = 3(1)^2 - 9(1) + 6$$

$$y = 3 - 9 + 6 = 0$$

In this situation, the value of x is more than y. If you substitute 5 for x, the value of y is:

$$y = 3(5)^2 - 9(5) + 6$$

$$y = 75 - 45 + 6 = 36$$

In this situation, the value of x is less than y. Therefore, using the information given, it cannot be determined which quantity is greater.

6. $\frac{4}{3}$: Common denominators must be used. The LCD is 15, and $\frac{2}{5} = \frac{6}{15}$. Therefore, $\frac{14}{15} + \frac{6}{15} = \frac{20}{15}$, and in lowest terms, the answer is $\frac{4}{3}$. A common factor of 5 was divided out of both the numerator and denominator.

7. 21: Gina answered 60% of 35 questions correctly; 60% can be expressed as the decimal 0.60. Therefore, she answered $0.60 \times 35 = 21$ questions correctly.

8. $\frac{2}{13}$: There is an equal chance of drawing a king or a queen. There are four kings and four queens so the probability of drawing either is:

$$\frac{4}{52} + \frac{4}{52} = \frac{8}{52} = \frac{2}{13}$$

9. 3:13: There were 48 total bags of apples sold. If 9 bags were Granny Smith and the rest were Red Delicious, then $48 - 9 = 39$ bags were Red Delicious. Therefore, the ratio of Granny Smith to Red Delicious is 9:39. Then it can be reduced to 3:13.

10. $9m^2$: The area of the shaded region is calculated in a few steps. First, the area of the rectangle is found using the formula:

$$A = length \times width = 6\text{ m} \times 2\text{ m} = 12\text{ m}^2$$

Second, the area of the triangle is found using the formula:

$$A = \frac{1}{2} \times base \times height = \frac{1}{2} \times 3\text{ m} \times 2\text{ m} = 3\text{ m}^2$$

The last step is to take the rectangle area and subtract the triangle area. The area of the shaded region is:

$$A = 12\text{ m}^2 - 3\text{ m}^2 = 9\text{ m}^2$$

11. A: First, the distributive property must be used on the left side. This results in:

$$3x + 6 = 14x - 5$$

The addition principle is then used to add 5 to both sides, and then to subtract $3x$ from both sides, resulting in $11 = 11x$. Finally, the multiplication principle is used to divide each side by 11. Therefore, $x = 1$ is the solution.

12. C: First, the variables have to be defined. Let x be the first integer; therefore, $x + 1$ is the second integer. This is a two-step problem. The sum of three times the first and two less than the second is translated into the following expression:

$$3x + (x + 1 - 2)$$

This expression is set equal to 411 to obtain:

$$3x + (x + 1 - 2) = 411$$

The left-hand side is simplified to obtain $4x - 1 = 411$. The addition and multiplication properties are used to solve for x. First, add 1 to both sides and then divide both sides by 4 to obtain $x = 103$. The next consecutive integer is 104.

13. A: Rounding can be used to find the best approximation. All of the values can be rounded to the nearest thousand. 15,412 SUVs can be rounded to 15,000. 25,815 station wagons can be rounded to 26,000. 50,412 sedans can be rounded to 50,000. 8,123 trucks can be rounded to 8,000. Finally, 18,312 hybrids can be rounded to 18,000. The sum of the rounded values is 117,000, which is closest to 120,000.

14. A: Let x be the unknown, the number of hours Erin can work. We know Katie works $2x$, and the sum of all hours is less than 21. Therefore, $x + 2x < 21$, which simplifies into $3x < 21$. Solving this results in the inequality $x < 7$ after dividing both sides by 3. Therefore, Erin can work less than 7 hours.

15. D: The volume for a cylinder is found by using the formula:

$$V = \pi r^2 h = \pi (2 \text{ in})^2 \times 3.5 \text{ in} = 43.96 \text{ in}^3$$

16. A and C: The chart is a bar chart showing how many men and women prefer each genre of movies. The dark gray bars represent the number of women, while the light gray bars represent the number of men. The light gray bars are higher and represent more men than women for the genres of Comedy and Action.

17. A, E, and G: The mode for a set of data is the value that occurs the most. The grades that appear the most are 65, 80, and 95. They are the only values that repeat in the set.

18. A, D, E, and G: The factors of 54 include those numbers that can be multiplied together to obtain a product of 54. They are 1, 2, 3, 6, 9, 18, 27, and 54.

19. C, E, and F: The slope-intercept form is $y = mx + b$ where b is the y-intercept, and m is the slope. The slope of the line can be found using rise over run or by substituting two points on the line into the slope formula:

$$m = \frac{(y_2 - y_1)}{(x_2 - x_1)}$$

The rise of the line is 2, and the run is 1. Therefore, the slope is $\frac{2}{1} = 2$. The y-intercept is the point where $x = 0$; the y-intercept of this line is −6. Therefore, the equation for the line is:

$$y = 2x - 6$$

20. A, C, and E: The first step is to substitute a_1 for a_n and a_2 for a_{n+1}. The formula then becomes $a_2 = 2a_1 + 3$. Then, 4 can be substituted for a_1:

$$a_2 = 2(4) + 3 = 11$$

This is one value for a_n. The steps can be repeated to find the other values for a_n which would include 25 and 53.

Analytical Writing

Articulating Ideas Clearly and Effectively

Possessives

Possessive forms indicate possession, i.e. that something belongs to or is owned by someone or something. As such, the most common parts of speech to be used in possessive form are adjectives, nouns, and pronouns. The rule for correctly spelling/punctuating possessive nouns and proper nouns is with -*'s,* like "the woman's briefcase" or "Frank's hat." With possessive adjectives, however, apostrophes are not used: these include *my, your, his, her, its, our,* and *their,* like "my book," "your friend," "his car," "her house," "its contents," "our family," or "their property." Possessive pronouns include *mine, yours, his, hers, its, ours,* and *theirs.* These also have no apostrophes. The difference is that possessive adjectives take direct objects, whereas possessive pronouns replace them. For example, instead of using two possessive adjectives in a row, as in "I forgot my book, so Blanca let me use her book," which reads monotonously, replacing the second one with a possessive pronoun reads better: "I forgot my book, so Blanca let me use hers."

Pronouns

There are three **pronoun** cases: **subjective case**, **objective case**, and **possessive case**. Pronouns as subjects are pronouns that replace the subject of the sentence, such as *I, you, he, she, it, we, they* and *who.* Pronouns as objects replace the object of the sentence, such as *me, you, him, her, it, us, them,* and *whom.* Pronouns that show possession are *mine, yours, hers, its, ours, theirs,* and *whose.* The following are examples of different pronoun cases:

- **Subject pronoun**: *She* ate the cake for her birthday. *I* saw the movie.
- **Object pronoun**: You gave *me* the card last weekend. She gave the picture to *him.*
- **Possessive pronoun**: That bracelet you found yesterday is *mine. His* name was Casey.

Adjectives

Adjectives are descriptive words that modify nouns or pronouns. They may occur before or after the nouns or pronouns they modify in sentences. For example, in "This is a big house," *big* is an adjective modifying or describing the noun *house.* In "This house is big," the adjective is at the end of the sentence rather than preceding the noun it modifies.

A rule of punctuation that applies to adjectives is to separate a series of adjectives with commas. For example, "Their home was a large, rambling, old, white, two-story house." A comma should never separate the last adjective from the noun, though.

Adverbs

Whereas adjectives modify and describe nouns or pronouns, **adverbs** modify and describe adjectives, verbs, or other adverbs. Adverbs can be thought of as answers to questions in that they describe when, where, how, how often, how much, or to what extent.

Many (but not all) adjectives can be converted to adverbs by adding *–ly.* For example, in "She is a quick learner," *quick* is an adjective modifying *learner.* In "She learns quickly," *quickly* is an adverb modifying *learns.* One exception is *fast. Fast* is an adjective in "She is a fast learner." However, *–ly* is never added to the word *fast,* it retains the same form as an adverb in "She learns fast."

Verbs

A verb is a word or phrase that expresses action, feeling, or state of being. Verbs explain what their subject is *doing*. Three different types of verbs used in a sentence are action verbs, linking verbs, and helping verbs.

Action verbs show a physical or mental action. Some examples of action verbs are *play, type, jump, write, examine, study, invent, develop,* and *taste*. The following example uses an action verb:

> Kat *imagines* that she is a mermaid in the ocean.

The verb *imagines* explains what Kat is doing: she is imagining being a mermaid.

Linking verbs connect the subject to the predicate without expressing an action. The following sentence shows an example of a linking verb:

> The mango *tastes* sweet.

The verb *tastes* is a linking verb. The mango doesn't *do* the tasting, but the word *taste* links the mango to its predicate, sweet. Most linking verbs can also be used as action verbs, such as *smell, taste, look, seem, grow,* and *sound*. Saying something *is* something else is also an example of a linking verb. For example, if we were to say, "Peaches is a dog," the verb *is* would be a linking verb in this sentence, since it links the subject to its predicate.

Helping verbs are verbs that help the main verb in a sentence. Examples of helping verbs are *be, am, is, was, have, has, do, did, can, could, may, might, should,* and *must*, among others. The following are examples of helping verbs:

> Jessica *is* planning a trip to Hawaii.
>
> Brenda *does* not like camping.
>
> Xavier *should* go to the dance tonight.

Notice that after each of these helping verbs is the main verb of the sentence: *planning, like,* and *go.* Helping verbs usually show an aspect of time.

Transitional Words and Phrases

In writing, some sentences naturally lead to others, whereas in other cases, a new sentence expresses a new idea. **Transitional phrases** connect sentences and the ideas they convey, which makes the writing coherent. Transitional language also guides the reader from one thought to the next. For example, when pointing out an objection to the previous idea, starting a sentence with "However," "But," or "On the other hand" is transitional. When adding another idea or detail, writers use "Also," "In addition," "Furthermore," "Further," "Moreover," "Not only," etc. Readers have difficulty perceiving connections between ideas without such transitional wording.

Subject-Verb Agreement

Lack of **subject-verb agreement** is a common grammatical error. One of the most common instances is when people use a series of nouns as a compound subject with a singular instead of a plural verb. Here is an example:

> Identifying the best books, locating the sellers with the lowest prices, and paying for them *is* difficult.

Instead of saying "*are* difficult." Additionally, when a sentence subject is compound, the verb is plural:

> He and his cousins *were* at the reunion.

However, if the conjunction connecting two or more singular nouns or pronouns is "or" or "nor," the verb must be singular to agree:

> That pen or another one like it is in the desk drawer.

If a **compound subject** includes both a singular noun and a plural one, and they are connected by "or" or "nor," the verb must agree with the subject closest to the verb: "Sally or her sisters go jogging daily"; but "Her sisters or Sally goes jogging daily."

Simply put, **singular subjects** require singular verbs and **plural subjects** require plural verbs. A common source of agreement errors is not identifying the sentence subject correctly. For example, people often write sentences incorrectly like, "The group of students *were* complaining about the test." The subject is not the plural "students" but the singular "group." Therefore, the correct sentence should read, "The group of students *was* complaining about the test." The inverse also applies, for example, in this incorrect sentence: "The facts in that complicated court case *is* open to question." The subject of the sentence is not the singular "case" but the plural "facts." Hence the sentence would correctly be written: "The facts in that complicated court case *are* open to question." New writers should not be misled by the distance between the subject and verb, especially when another noun with a different number intervenes as in these examples. The verb must agree with the subject, not the noun closest to it.

Pronoun-Antecedent Agreement

Pronouns within a sentence must refer specifically to one noun, known as the **antecedent**. Sometimes, if there are multiple nouns within a sentence, it may be difficult to ascertain which noun belongs to the pronoun. It's important that the pronouns always clearly reference the nouns in the sentence so as not to confuse the reader. Here's an example of an unclear pronoun reference:

> After Catherine cut Libby's hair, David bought her some lunch.

The pronoun in the examples above is *her*. The pronoun could either be referring to *Catherine* or *Libby*. Here are some ways to write the above sentence with a clear pronoun reference:

> After Catherine cut Libby's hair, David bought Libby some lunch.

> David bought Libby some lunch after Catherine cut Libby's hair.

But many times, the pronoun will clearly refer to its antecedent, like the following:

> After David cut Catherine's hair, he bought her some lunch.

Support Ideas with Relevant Reasons and Examples

Selecting the most relevant material to support a written text is a necessity in producing quality writing and for the credibility of an author. Arguments lacking in reasons or examples won't work in persuading the audience later on, because their hearts have not been pulled. Using examples to support ideas also gives the writing rhetorical effects such as **pathos** (emotion), **logos** (logic), or **ethos** (credibility), all three of which are necessary for a successful text. There are a few steps in determining how to support an idea:

Types of Examples

Think about the audience. Are they indifferent? Use a personal story or example in order stir empathy. Are they resistant? Use logical reasoning based in factual evidence so they will be convinced. Personal stories or testimonials, statistics, or documentary evidence are various types of examples that one can use for their writing.

Source and Validity

For libraries or online sources, make sure the research you find is within a credible, scholarly journal and is peer-reviewed. Peer-reviewed sources are sources that have been reviewed by other experts in the field. If scholarly journals do not have the information you are looking for, be sure to research the author, date, and website of the source you find, identifying any credibility issues in the process. Sources must also be relevant by being up-to-date, especially those within the science or technology fields.

For example, let's say we want to talk about pesticides and the collapse of bees. An argument without relevant examples would look like this:

> With the use of the world's most popular pesticides, bees are becoming extinct. This is also causing ecological devastation. We must do something soon about the bee population, or else we will chase bees to extinction and lose valuable resources as a result.

Here is the same argument with examples. The added examples are in italics:

> With the use of the world's most popular pesticides, bees are becoming extinct. *Beekeepers have reported losing 55 to 95 percent of their colony in just two short years due to toxic poisoning.* This is also causing ecological devastation. *Bees are known for pollinating more than two-thirds of the world's most essential crops.* We must do something soon about the bee population, or else we will chase bees to extinction and lose valuable resources as a result.

Adding examples to the above argument brings life to the bees—they are living, dying, pollinating—and readers feel more compelled to act as a result of adding relevant examples to the argument.

Examining Claims and Evidence

When authors write text for the purpose of persuading others to agree with them, they assume a position with the subject matter about which they are writing. Rather than presenting information objectively, the author treats the subject matter subjectively so that the information presented supports their position. In their argumentation, the author presents information that refutes or weakens opposing positions. Another technique authors use in persuasive writing is to anticipate arguments against the position. When students learn to read subjectively, they gain experience with the concept of persuasion in writing and learn to identify positions taken by authors. This enhances their reading comprehension and develops their skills for identifying pro and con arguments and biases.

There are five main parts of the **classical argument** that writers employ in a well-designed stance:

- **Introduction**: In the introduction to a classical argument, the author establishes goodwill and rapport with the reading audience, warms up the readers, and states the thesis or general theme of the argument.
- **Narration**: In the narration portion, the author gives a summary of pertinent background information, informs the readers of anything they need to know regarding the circumstances and environment surrounding and/or stimulating the argument, and establishes what is at risk or the stakes in the issue or topic. Literature reviews are common examples of narrations in academic writing.
- **Confirmation**: The confirmation states all claims supporting the thesis and furnishes evidence for each claim, arranging this material in logical order—e.g. from most obvious to most subtle or strongest to weakest.
- **Refutation and Concession**: The refutation and concession discuss opposing views and anticipate reader objections without weakening the thesis, yet permitting as many oppositions as possible.
- **Summation**: The summation strengthens the argument while summarizing it, supplying a strong conclusion and showing readers the superiority of the author's solution.

Introduction

A classical argument's introduction must pique reader interest, get readers to perceive the author as a writer, and establish the author's position. Shocking statistics, new ways of restating issues, or quotations or anecdotes focusing the text can pique reader interest. Personal statements, parallel instances, or analogies can also begin introductions—so can bold thesis statements if the author believes readers will agree. Word choice is also important for establishing author image with readers.

The introduction should typically narrow down to a clear, sound thesis statement. If readers cannot locate one sentence in the introduction explicitly stating the writer's position or the point they support, the writer probably has not refined the introduction sufficiently.

Narration and Confirmation

The narration part of a classical argument should create a context for the argument by explaining the issue to which the argument is responding, and by supplying any background information that influences the issue. Readers should understand the issues, alternatives, and stakes in the argument by the end of the narration to enable them to evaluate the author's claims equitably. The confirmation part of the classical argument enables the author to explain why they believe in the argument's thesis. The author builds a chain of reasoning by developing several individual supporting claims and explaining why that evidence supports each claim and also supports the overall thesis of the argument.

Refutation and Concession and Summation

The classical argument is the model for argumentative/persuasive writing, so authors often use it to establish, promote, and defend their positions. In the refutation aspect of the refutation and concession part of the argument, authors disarm reader opposition by anticipating and answering their possible objections, persuading them to accept the author's viewpoint. In the concession aspect, authors can concede those opposing viewpoints with which they agree. This can avoid weakening the author's thesis while establishing reader respect and goodwill for the author: all refutation and no concession can antagonize readers who disagree with the author's position. In the conclusion part of the classical argument, a less skilled writer might simply summarize or restate the thesis and related claims; however, this does not provide the argument with either momentum or closure. More skilled authors revisit the issues and the narration part of the argument, reminding readers of what is at stake.

Sustaining a Well-Focused, Coherent Discussion

Formal and Informal Language

Formal language is less personal than informal language. It is more "buttoned-up" and business-like, adhering to proper grammatical rules. It is used in professional or academic contexts, to convey respect or authority. For example, one would use formal language to write an informative or argumentative essay for school or to address a superior. Formal language avoids contractions, slang, colloquialisms, and first-person pronouns. Formal language uses sentences that are usually more complex and often in passive voice. Punctuation can differ as well. For example, *exclamation points (!)* are used to show strong emotion or can be used as an **interjection** but should be used sparingly in formal writing situations.

Informal language is often used when communicating with family members, friends, peers, and those known more personally. It is more casual, spontaneous, and forgiving in its conformity to grammatical rules and conventions. Informal language is used for personal emails and correspondence between coworkers or other familial relationships. The tone is more relaxed. In informal writing, slang, contractions, clichés, and the first- and second-person are often used.

Elements of the Writing Process

Skilled writers undergo a series of steps that comprise the writing process. The purpose of adhering to a structured approach to writing is to develop clear, meaningful, coherent work.

The stages are pre-writing or planning, organizing, drafting/writing, revising, and editing. Not every writer will necessarily follow all five stages for every project, but will judiciously employ the crucial components of the stages for most formal or important work. For example, a brief informal response to a short reading passage may not necessitate the need for significant organization after idea generation, but larger assignments and essays will likely mandate use of the full process.

Pre-Writing/Planning

Brainstorming

Before beginning the essay, read the prompt thoroughly and make sure you understand its expectations. Brainstorm as many ideas as you can think of that relate to the topic and list them or put them into a graphic organizer. Refer to this list as you organize your essay outline.

Freewriting

Like brainstorming, **freewriting** is another prewriting activity to help the writer generate ideas. This method involves setting a timer for two or three minutes and writing down all ideas that come to mind about the topic using complete sentences. Once time is up, writers should review the sentences to see what observations have been made and how these ideas might translate into a more unified direction for the topic. Even if sentences lack sense as a whole, freewriting is an excellent way to get ideas onto the page in the very beginning stages of writing. Using complete sentences can make this a bit more challenging than brainstorming, but overall it is a worthwhile exercise, as it may force the writer to come up with more complete thoughts about the topic.

Take the ideas you generated during pre-writing and organize them in the form of an outline.

Organizing

Although sometimes it is difficult to get going on the brainstorming or prewriting phase, once ideas start flowing, writers often find that they have amassed too many thoughts that will not make for a cohesive and unified essay. During the **organization** stage, writers should examine the generated ideas, hone in on the important ones central to their main idea, and arrange the points in a logical and effective manner. Writers may also determine that some of the ideas generated in the planning process need further elaboration, potentially necessitating the need for research to gather information to fill the gaps.

Once a writer has chosen their thesis and main argument, selected the most applicable details and evidence, and eliminated the "clutter," it is time to strategically organize the ideas. This is often accomplished with an outline.

Outlining

Outlines are organizational tools that arrange a piece of writing's main ideas and the evidence that supports them. After pre-writing, organize your ideas by topic, select the best ones, and put them into the outline. Be sure to include an introduction, main points, and a conclusion. Typically, it is a good idea to have three main points with at least two pieces of supporting evidence each. The following displays the format of an outline:

- I. Introduction
    1. Background
    2. Thesis statement
- II. Body
    1. Point A
        - a. Supporting evidence
        - b. Supporting evidence
    2. Point B
        - a. Supporting evidence
        - b. Supporting evidence
    3. Point C
        - a. Supporting evidence
        - b. Supporting evidence
- III. Conclusion
    1. Restatement of main points.
    2. Memorable ending.

Drafting/Writing

Now it comes time to actually write the essay. In this stage, writers should follow the outline they developed in the brainstorming process and try to incorporate the useful sentences penned in the freewriting exercise. The main goal of this phase is to put all the thoughts together in cohesive sentences and paragraphs.

It is helpful for writers to remember that their work here does not have to be perfect. This process is often referred to as **drafting** because writers are just creating a rough draft of their work. Because of this, writers should avoid getting bogged down on the small details.

Referencing Sources

Anytime a writer quotes or paraphrases another text, they will need to include a citation. A **citation** is a short description of the work that a quote or information came from. The style manual your teacher wants you to follow will dictate exactly how to format that citation. For example, this is how one would cite a book according to the APA manual of style:

- *Format*: Last name, First initial, Middle initial. (Year Published) *Book Title*. City, State: Publisher.
- *Example*: Sampson, M. R. (1989). *Diaries from an Alien Invasion. Springfield, IL*: Campbell Press.

Revising

Revising involves going back over a piece of writing and improving it. Try to read your essay from the perspective of a potential reader to ensure that it makes sense. When revising, check that the main points are clearly stated, logically organized, and directly supported by the sub-points. Remove unnecessary details that do not contribute to the argument.

The main goal of the revision phase is to improve the essay's flow, cohesiveness, readability, and focus. For example, an essay will make a less persuasive argument if the various pieces of evidence are scattered and presented illogically or clouded with unnecessary thought. Therefore, writers should consider their essay's structure and organization, ensuring that there are smooth transitions between sentences and paragraphs. There should be a discernable introduction and conclusion as well, as these crucial components of an essay provide readers with a blueprint to follow.

Additionally, if the writer includes copious details that do little to enhance the argument, they may actually distract readers from focusing on the main ideas and detract from the strength of their work. The ultimate goal is to retain the purpose or focus of the essay and provide a reader-friendly experience. Because of this, writers often need to delete parts of their essay to improve its flow and focus. Removing sentences, entire paragraphs, or large chunks of writing can be one of the toughest parts of the writing process because it is difficult to part with work one has done. However, ultimately, these types of cuts can significantly improve one's essay.

Lastly, writers should consider their voice and word choice. The voice should be consistent throughout and maintain a balance between an authoritative and warm style, to both inform and engage readers. One way to alter voice is through word choice. Writers should consider changing weak verbs to stronger ones and selecting more precise language in areas where wording is vague. In some cases, it is useful to modify sentence beginnings or to combine or split up sentences to provide a more varied sentence structure.

Editing

Rather than focusing on content (as is the aim in the revising stage), the **editing** phase is all about the mechanics of the essay: the syntax, word choice, and grammar. This can be considered the **proofreading** stage. Successful editing is what sets apart a messy essay from a polished document.

Look for the following types of errors when checking over your work:

1. Spelling
2. Tense usage
3. Punctuation and capitalization
4. Unclear, confusing, or incomplete sentences

5. Subject/verb and noun/pronoun agreement

One of the most effective ways of identifying grammatical errors, awkward phrases, or unclear sentences is to read the essay out loud. Listening to one's own work can help move the writer from simply the author to the reader.

During the editing phase, it's also important to ensure the essay follows the correct formatting and citation rules as dictated by the assignment.

Recursive Writing Process

While the writing process may have specific steps, the good news is that the process is **recursive**, meaning the steps need not be completed in a particular order. Many writers find that they complete steps at the same time such as drafting and revising, where the writing and rearranging of ideas occur simultaneously or in very close order. Similarly, a writer may find that a particular section of a draft needs more development, and will go back to the prewriting stage to generate new ideas. The steps can be repeated at any time, and the more these steps of the recursive writing process are employed, the better the final product will be.

Practice Makes Prepared Writers

Like any other useful skill, writing only improves with practice. While writing may come more easily to some than others, it is still a skill to be honed and improved. Regardless of a person's natural abilities, there is always room for growth in writing. Practicing the basic skills of writing can aid in preparations for the exam.

One way to build vocabulary and enhance exposure to the written word is through reading. This can be through reading books, but reading of any materials such as newspapers, magazines, and even social media count towards practice with the written word. This also helps to enhance critical reading and thinking skills, through analysis of the ideas and concepts read. Think of each new reading experience as a chance to sharpen these skills.

Developing a Well-Organized Paragraph

A **paragraph** is a series of connected and related sentences addressing one topic. Writing good paragraphs benefits writers by helping them to stay on target while drafting and revising their work. It benefits readers by helping them to follow the writing more easily. Regardless of how brilliant their ideas may be, writers who do not present them in organized ways will fail to engage readers—and fail to accomplish their writing goals. A fundamental rule for paragraphing is to confine each paragraph to a single idea. When writers find themselves transitioning to a new idea, they should start a new paragraph. However, a paragraph can include several pieces of evidence supporting its single idea; and it can include several points if they are all related to the overall paragraph topic. When writers find each point becoming lengthy, they may choose instead to devote a separate paragraph to every point and elaborate upon each more fully.

An effective paragraph should have these elements:

- **Unity**: One major discussion point or focus should occupy the whole paragraph from beginning to end.
- **Coherence**: For readers to understand a paragraph, it must be coherent. Two components of coherence are logical and verbal bridges. In logical bridges, the writer may write consecutive sentences with parallel structure or carry an idea over across sentences. In verbal bridges, writers may repeat key words across sentences.
- A **topic sentence**: The paragraph should have a sentence that generally identifies the paragraph's thesis or main idea.
- Sufficient **development**: To develop a paragraph, writers can use the following techniques after stating their topic sentence:
    - Define terms
    - Cite data
    - Use illustrations, anecdotes, and examples
    - Evaluate causes and effects
    - Analyze the topic
    - Explain the topic using chronological order

A **topic sentence** identifies the main idea of the paragraph. Some are explicit, some implicit. The topic sentence can appear anywhere in the paragraph. However, many experts advise beginning writers to place each paragraph topic sentence at or near the beginning of its paragraph to ensure that their readers understand what the topic of each paragraph is. Even without having written an explicit topic sentence, the writer should still be able to summarize readily what subject matter each paragraph addresses. The writer must then fully develop the topic that is introduced or identified in the topic sentence. Depending on what the writer's purpose is, they may use different methods for developing each paragraph.

Two main steps in the process of organizing paragraphs and essays should both be completed after determining the writing's main point, while the writer is planning or outlining the work. The initial step is to give an order to the topics addressed in each paragraph. Writers must have logical reasons for putting one paragraph first, another second, etc. The second step is to sequence the sentences in each paragraph. As with the first step, writers must have logical reasons for the order of sentences. Sometimes the work's main point obviously indicates a specific order.

Topic Sentences

To be effective, a topic sentence should be concise so that readers get its point without losing the meaning among too many words. As an example, in *Only Yesterday: An Informal History of the 1920s* (1931), author Frederick Lewis Allen's topic sentence introduces his paragraph describing the 1929 stock market crash: "The Bull Market was dead." This example illustrates the criteria of conciseness and brevity. It is also a strong sentence, expressed clearly and unambiguously. The topic sentence also introduces the paragraph, alerting the reader's attention to the main idea of the paragraph and the subject matter that follows the topic sentence.

Experts often recommend opening a paragraph with the topic sentences to enable the reader to realize the main point of the paragraph immediately. Application letters for jobs and university admissions also benefit from opening with topic sentences. However, positioning the topic sentence at the end of a

paragraph is more logical when the paragraph identifies a number of specific details that accumulate evidence and then culminates with a generalization. While paragraphs with extremely obvious main ideas need no topic sentences, more often—and particularly for students learning to write—the topic sentence is the most important sentence in the paragraph. It not only communicates the main idea quickly to readers; it also helps writers produce and control information.

Controlling the Elements of Standard Written English

Conventions of Standard English Spelling

Homophones

Homophones are words that have different meanings and spellings but sound the same. These can be confusing for English Language Learners (ELLs) and beginning students, but even native English-speaking adults can find them problematic unless informed by context. Whereas listeners must rely entirely on context to **differentiate** spoken homophone meanings, readers with good spelling knowledge have a distinct advantage since homophones are spelled differently.

For instance, *their* means belonging to them; *there* indicates location; and *they're* is a contraction of *they are*; despite different meanings, they all sound the same. *Lacks* can be a plural noun or a present-tense, third-person singular verb; either way it refers to absence—*deficiencies* as a plural noun, and *is deficient in* as a verb. But *lax* is an adjective that means loose, slack, relaxed, uncontrolled, or negligent. These two spellings, derivations, and meanings are completely different. With speech, listeners cannot know spelling and must use context; but with print, readers with spelling knowledge can differentiate them with or without context.

Homonyms, Homophones, and Homographs

Homophones are words that sound the same in speech, but have different spellings and meanings. For example, *to, too,* and *two* all sound alike, but have three different spellings and meanings. Homophones with different spellings are also called **heterographs**. **Homographs** are words that are spelled identically, but have different meanings. If they also have different pronunciations, they are **heteronyms**. For instance, *tear* pronounced one way means a drop of liquid formed by the eye; pronounced another way, it means to rip. Homophones that are also homographs are **homonyms**. For example, *bark* can mean the outside of a tree or a dog's vocalization; both meanings have the same spelling. *Stalk* can mean a plant stem or to pursue and/or harass somebody; these are spelled and pronounced the same. *Rose* can mean a flower or the past tense of *rise*. Many non-linguists confuse things by using "homonym" to mean sets of words that are homophones but not homographs, and also those that are homographs but not homophones.

The word *row* can mean to use oars to propel a boat; a linear arrangement of objects or print; or an argument. It is pronounced the same with the first two meanings, but differently with the third. Because it is spelled identically regardless, all three meanings are homographs. However, the two meanings pronounced the same are homophones, whereas the one with the different pronunciation is a heteronym. By contrast, the word *read* means to peruse language, whereas the word *reed* refers to a marsh plant.

Because these are pronounced the same way, they are homophones; because they are spelled differently, they are heterographs. Homonyms are both homophones and homographs—pronounced and spelled identically, but with different meanings. One distinction between homonyms is of those with separate, unrelated etymologies, called "true" homonyms, e.g. *skate* meaning a fish or *skate* meaning to glide over ice/water. Those with common origins are called **polysemes** or **polysemous homonyms**, e.g. the *mouth* of an animal/human or of a river.

Irregular Plurals

While many words in English can become plural by adding –s or –es to the end, there are some words that have **irregular plural forms.** One type includes words that are spelled the same whether they are singular or plural, such as deer, fish, salmon, trout, sheep, moose, offspring, species, aircraft, etc. The spelling rule for making these words plural is simple: they do not change. Other irregular English plurals change form based on vowel shifts, linguistic mutations, or grammatical and spelling conventions from their languages of origin, like Latin or German. Some examples include *child* and *children; die* and *dice; foot* and *feet; goose* and *geese; louse* and *lice; man* and *men; mouse* and *mice; ox* and *oxen; person* and *people; tooth* and *teeth;* and *woman* and *women.*

Contractions

Contractions are formed by joining two words together, omitting one or more letters from one of the component words, and replacing the omitted letter(s) with an apostrophe. An obvious yet often forgotten rule for spelling contractions is to place the apostrophe where the letters were omitted. For example, didn't is a contraction of did not; therefore, the apostrophe replaces the "o" that is omitted from the "not." Another common error is confusing contractions with possessives because both include apostrophes, e.g. spelling the possessive *its* as "it's," which is a contraction of "it is"; spelling the possessive *their* as "they're," a contraction of "they are"; spelling the possessive *whose* as "who's," a contraction of "who is"; or spelling the possessive *your* as "you're," a contraction of "you are."

Frequently Misspelled Words

One source of spelling errors is not knowing whether to drop the final letter *e* from a word when its form is changed; some words retain the final *e* when another syllable is added while others lose it. For example, *true* becomes *truly; argue* becomes *arguing; come* becomes *coming; write* becomes *writing;* and *judge* becomes *judging.* In these examples, the final *e* is dropped before adding the ending. But *severe* becomes *severely; complete* becomes *completely; sincere* becomes *sincerely; argue* becomes *argued;* and *care* becomes *careful.* In these instances, the final *e* is retained before adding the ending. Note that some words, like argue in these examples, drops the final e when the –ing ending is added to indicate the participial form, but the regular past tense form keeps the e and adds a –d to make it argued.

Other commonly misspelled English words are those containing the vowel combinations ei and ie. Many people confuse these two. Some examples of words with the ei combination include:

> *ceiling, conceive, leisure, receive, weird, their, either, foreign, sovereign, neither, neighbors, seize, forfeit, counterfeit, height, weight, protein,* and *freight*

Words with *ie* include *piece, believe, chief, field, friend, grief, relief, mischief, siege, niece, priest, fierce, pierce, achieve, retrieve, hygiene, science,* and *diesel.* A rule that also functions as a mnemonic device is "I before E except after C, or when sounded like A as in 'neighbor' or 'weigh'." However, it is obvious from the list above that many exceptions exist.

People often misspell certain words by confusing whether they have the vowel a, e, or i. For example, in the following correctly spelled words, the vowel in boldface is the one people typically get wrong by substituting one of the others for it:

> cem**e**tery, quant**i**ties, ben**e**fit, priv**i**lege, unpleas**a**nt, sep**a**rate, independ**e**nt, excell**e**nt, cat**e**gories, indispens**a**ble, and irrelev**a**nt

Some words with final syllables that sound the same when spoken but are spelled differently include *unpleasant, independent, excellent,* and *irrelevant.* Another source of misspelling is whether or not to double consonants when adding suffixes. For example, double the last consonant before –ed and –ing endings in controlled, beginning, forgetting, admitted, occurred, referred, and hopping; but do not double before the suffix in *shining, poured, sweating, loving, hating, smiling,* and *hoping.*

One final example of common misspellings involves either the failure to include silent letters or the converse of adding extraneous letters. If a letter is not pronounced in speech, it is easy to leave it out in writing. For example, some people omit the silent *u* in *g**u**arantee,* overlook the first *r* in *su**r**prise,* leave out the *z* in *reali**z**e,* fail to double the *m* in *recom**m**end,* leave out the middle *i* from *asp**i**rin,* and exclude the *p* from *tem**p**erature.* The converse error, adding extra letters, is common in words like *until* by adding a second *l* at the end; or by inserting a superfluous syllabic *a* or *e* in the middle of *athletic,* reproducing a common mispronunciation.

Conventions of Standard English Punctuation

Rules of Capitalization

The first word of any document, and of each new sentence, is capitalized. Proper nouns, like names and adjectives derived from proper nouns, should also be capitalized. Here are some examples:

- Grand Canyon
- Pacific Palisades
- Golden Gate Bridge
- Freudian slip
- Shakespearian, Spenserian, or Petrarchan sonnet
- Irish song

Some exceptions are adjectives, originally derived from proper nouns, which through time and usage are no longer capitalized, like *quixotic, herculean,* or *draconian.* Capitals draw attention to specific instances of people, places, and things.

Some categories that should be capitalized include the following:

- Brand names
- Companies
- Months
- Governmental divisions or agencies
- Historical eras
- Major historical events
- Holidays
- Institutions
- Famous buildings
- Ships and other manmade constructions
- Natural and manmade landmarks
- Territories
- Nicknames
- Organizations
- Planets
- Nationalities
- Tribes
- Religions
- Names of religious deities

- Roads
- Special occasions, like the Cannes Film Festival or the Olympic Games

Exceptions

Related to American government, capitalize the noun Congress but not the related adjective congressional. Capitalize the noun U.S. Constitution, but not the related adjective constitutional. Many experts advise leaving the adjectives federal and state in lowercase, as in federal regulations or state water board, and only capitalizing these when they are parts of official titles or names, like Federal Communications Commission or State Water Resources Control Board. While the names of the other planets in the solar system are capitalized as names, Earth is more often capitalized only when being described specifically as a planet, like Earth's orbit, but lowercase otherwise since it is used not only as a proper noun but also to mean *land, ground, soil,* etc. While the name of the Bible should be capitalized, the adjective biblical should not. Regarding biblical subjects, the words heaven, hell, devil, and satanic should not be capitalized. Although race names like Caucasian, African-American, Navajo, Eskimo, East Indian, etc. are capitalized, white and black as races are not.

Names of animal species or breeds are not capitalized unless they include a proper noun. Then, only the proper noun is capitalized. Antelope, black bear, and yellow-bellied sapsucker are not capitalized. However, Bengal tiger, German shepherd, Australian shepherd, French poodle, and Russian blue cat are capitalized.

Other than planets, celestial bodies like the sun, moon, and stars are not capitalized. Medical conditions like tuberculosis or diabetes are lowercase; again, exceptions are proper nouns, like Epstein-Barr syndrome, Alzheimer's disease, and Down syndrome. Seasons and related terms like winter solstice or autumnal equinox are lowercase. Plants, including fruits and vegetables, like poinsettia, celery, or avocados, are not capitalized unless they include proper names, like Douglas fir, Jerusalem artichoke, Damson plums, or Golden Delicious apples.

Titles and Names

When official titles precede names, they should be capitalized, except when there is a comma between the title and name. But if a title follows or replaces a name, it should not be capitalized. For example, "the president" without a name is not capitalized, as in "The president addressed Congress." But with a name it is capitalized, like "President Obama addressed Congress." Or, "Chair of the Board Janet Yellen was appointed by President Obama." One exception is that some publishers and writers nevertheless capitalize President, Queen, Pope, etc., when these are not accompanied by names to show respect for these high offices. However, many writers in America object to this practice for violating democratic principles of equality. Occupations before full names are not capitalized, like owner Mark Cuban, director Martin Scorsese, or coach Roger McDowell.

Some universal rules for capitalization in composition titles include capitalizing the following:

- The first and last words of the title
- Forms of the verb *to be* and all other verbs
- Pronouns
- The word *not*

Universal rules for NOT capitalizing include the articles *the, a,* or *an;* the conjunctions *and, or,* or *nor,* and the preposition *to,* or *to* as part of the infinitive form of a verb. The exception to all of these is UNLESS any of them is the first or last word in the title, in which case they are capitalized. Other words are subject to differences of opinion and differences among various stylebooks or methods. These include *as, but, if,* and *or,* which some capitalize and others do not. Some authorities say no preposition should ever be capitalized; some say prepositions five or more letters long should be capitalized. The *Associated Press Stylebook* advises capitalizing prepositions longer than three letters (like *about, across,* or *with*).

Ellipses

Ellipses (. . .) signal omitted text when quoting. Some writers also use them to show a thought trailing off, but this should not be overused outside of dialogue. An example of an ellipsis would be if someone is quoting a phrase out of a professional source but wants to omit part of the phrase that isn't needed: "Dr. Skim's analysis of pollen inside the body is clearly a myth . . . that speaks to the environmental guilt of our society."

Commas

Commas separate words or phrases in a series of three or more. The **Oxford comma** is the last comma in a series. Many people omit this last comma, but many times it causes confusion. Here is an example:

> I love my sisters, the Queen of England and Madonna.

This example without the comma implies that the "Queen of England and Madonna" are the speaker's sisters. However, if the speaker was trying to say that they love their sisters, the Queen of England, as well as Madonna, there should be a comma after "Queen of England" to signify this.

Commas also separate two coordinate adjectives ("big, heavy dog") but not cumulative ones, which should be arranged in a particular order for them to make sense ("beautiful ancient ruins").

A comma ends the first of two independent clauses connected by conjunctions. Here is an example:

> I ate a bowl of tomato soup, and I was hungry very shortly after.

Here are some brief rules for commas:

- Commas follow introductory words like *however, furthermore, well, why,* and *actually,* among others.
- Commas go between city and state: Houston, Texas.
- If using a comma between a surname and Jr. or Sr. or a degree like M.D., also follow the whole name with a comma: "Martin Luther King, Jr., wrote that."
- A comma follows a dependent clause beginning a sentence: "Although she was very small, . . ."
- Nonessential modifying words/phrases/clauses are enclosed by commas: "Wendy, who is Peter's sister, closed the window."
- Commas introduce or interrupt direct quotations: "She said, 'I hate him.' 'Why,' I asked, 'do you hate him?'"

Semicolons

Semicolons are used to connect two independent clauses, but should never be used in the place of a comma. They can replace periods between two closely connected sentences: "Call back tomorrow; it can wait until then." When writing items in a series and one or more of them contains internal commas, separate them with semicolons, like the following:

> People came from Springfield, Illinois; Alamo, Tennessee; Moscow, Idaho; and other locations.

Hyphens

Here are some rules concerning hyphens:

- Compound adjectives like state-of-the-art or off-campus are hyphenated.
- Original compound verbs and nouns are often hyphenated, like "throne-sat," "video-gamed," "no-meater."
- Adjectives ending in *–ly* are often hyphenated, like "family-owned" or "friendly-looking."
- "Five years old" is not hyphenated, but singular ages like "five-year-old" are.
- Hyphens can clarify. For example, in "stolen vehicle report," "stolen-vehicle report" clarifies that "stolen" modifies "vehicle," not "report."
- Compound numbers twenty-one through ninety-nine are spelled with hyphens.
- Prefixes before proper nouns/adjectives are hyphenated, like "mid-September" and "trans-Pacific."

Parentheses

Parentheses enclose information such as an aside or more clarifying information: "She ultimately replied (after deliberating for an hour) that she was undecided." They are also used to insert short, in-text definitions or acronyms: "His FBS (fasting blood sugar) was higher than normal." When parenthetical information ends the sentence, the period follows the parentheses: "We received new funds ($25,000)." Only put periods within parentheses if the whole sentence is inside them: "Look at this. (You'll be astonished.)" However, this can also be acceptable as a clause: "Look at this (you'll be astonished)." Although parentheses appear to be part of the sentence subject, they are not, and do not change subject-verb agreement: "Will (and his dog) was there."

Quotation Marks

Quotation marks are typically used when someone is quoting a direct word or phrase someone else writes or says. Additionally, quotation marks should be used for the titles of poems, short stories, songs, articles, chapters, and other shorter works. When quotations include punctuation, periods and commas should *always* be placed inside of the quotation marks.

When a quotation contains another quotation inside of it, the outer quotation should be enclosed in double quotation marks and the inner quotation should be enclosed in single quotation marks. For example: "Timmy was begging, 'Don't go! Don't leave!'" When using both double and single quotation marks, writers will find that many word-processing programs may automatically insert enough space between the single and double quotation marks to be visible for clearer reading. But if this is not the case, the writer should write/type them with enough space between to keep them from looking like three single quotation marks. Additionally, non-standard usages, terms used in an unusual fashion, and technical terms are often clarified by quotation marks. Here are some examples:

> My "friend," Dr. Sims, has been micromanaging me again.
>
> This way of extracting oil has been dubbed "fracking."

Apostrophes

One use of the **apostrophe** is followed by an *s* to indicate possession, like *Mrs. White's home* or *our neighbor's dog*. When using the *'s* after names or nouns that also end in the letter *s*, no single rule applies: some experts advise adding both the apostrophe and the *s,* like "the Jones's house," while others prefer using only the apostrophe and omitting the additional *s,* like "the Jones' house." The wisest expert advice is to pick one formula or the other and then apply it consistently. Newspapers and magazines often use *'s* after common nouns ending with *s,* but add only the apostrophe after proper nouns or names ending with *s.* One common error is to place the apostrophe before a name's final *s* instead of after it: "Ms. Hasting's book" is incorrect if the name is Ms. Hastings.

Plural nouns should not include apostrophes (e.g. "apostrophe's"). Exceptions are to clarify atypical plurals, like verbs used as nouns: "These are the do's and don'ts." Irregular plurals that do not end in *s* always take apostrophe-*s,* not *s*-apostrophe—a common error, as in "childrens' toys," which should be "children's toys." Compound nouns like mother-in-law, when they are singular and possessive, are followed by apostrophe-*s,* like "your mother-in-law's coat." When a compound noun is plural and possessive, the plural is formed before the apostrophe-*s,* like "your sisters-in-laws' coats." When two people named possess the same thing, use apostrophe-*s* after the second name only, like "Dennis and Pam's house."

Sentence Structures

Incomplete Sentences

Four types of incomplete sentences are sentence fragments, run-on sentences, subject-verb and/or pronoun-antecedent disagreement, and non-parallel structure.

Sentence fragments are caused by absent subjects, absent verbs, or dangling/uncompleted dependent clauses. Every sentence must have a subject and a verb to be complete. An example of a fragment is "Raining all night long," because there is no subject present. "It was raining all night long" is one correction. Another example of a sentence fragment is the second part in "Many scientists think in unusual ways. Einstein, for instance." The second phrase is a fragment because it has no verb. One correction is "Many scientists, like Einstein, think in unusual ways." Finally, look for "cliffhanger" words like *if, when, because,* or *although* that introduce dependent clauses, which cannot stand alone without an independent clause. For example, to correct the sentence fragment "If you get home early," add an independent clause: "If you get home early, we can go dancing."

Run-On Sentences

A **run-on sentence** combines two or more complete sentences without punctuating them correctly or separating them. For example, a run-on sentence caused by a lack of punctuation is the following:

> There is a malfunction in the computer system however there is nobody available right now who knows how to troubleshoot it.

One correction is, "There is a malfunction in the computer system; however, there is nobody available right now who knows how to troubleshoot it." Another is, "There is a malfunction in the computer system. However, there is nobody available right now who knows how to troubleshoot it."

An example of a comma splice of two sentences is the following:

> Jim decided not to take the bus, he walked home.

Replacing the comma with a period or a semicolon corrects this. Commas that try and separate two independent clauses without a conjunction are considered comma splices.

Parallel Sentence Structures

Parallel structure in a sentence matches the forms of sentence components. Any sentence containing more than one description or phrase should keep them consistent in wording and form. Readers can easily follow writers' ideas when they are written in parallel structure, making it an important element of correct sentence construction. For example, this sentence lacks parallelism: "Our coach is a skilled manager, a clever strategist, and works hard." The first two phrases are parallel, but the third is not. Correction: "Our coach is a skilled manager, a clever strategist, and a hard worker." Now all three phrases match in form. Here is another example:

> Fred intercepted the ball, escaped tacklers, and a touchdown was scored.

This is also non-parallel. Here is the sentence corrected:

> Fred intercepted the ball, escaped tacklers, and scored a touchdown.

Sentence Fluency

For fluent composition, writers must use a variety of sentence types and structures, and also ensure that they smoothly flow together when they are read. To accomplish this, they must first be able to identify fluent writing when they read it. This includes being able to distinguish among simple, compound, complex, and compound-complex sentences in text; to observe variations among sentence types, lengths, and beginnings; and to notice figurative language and understand how it augments sentence length and imparts musicality. Once students/writers recognize superior fluency, they should revise their own writing to be more readable and fluent. They must be able to apply acquired skills to revisions before being able to apply them to new drafts.

One strategy for revising writing to increase its sentence fluency is flipping sentences. This involves rearranging the word order in a sentence without deleting, changing, or adding any words. For example, the student or other writer who has written the sentence, "We went bicycling on Saturday" can revise it to, "On Saturday, we went bicycling." Another technique is using appositives. An **appositive** is a phrase or word that renames or identifies another adjacent word or phrase. Writers can revise for sentence fluency by inserting main phrases/words from one shorter sentence into another shorter sentence, combining them into one longer sentence, e.g. from "My cat Peanut is a gray and brown tabby. He loves hunting rats." to "My cat Peanut, a gray and brown tabby, loves hunting rats." Revisions can also connect shorter sentences by using conjunctions and commas and removing repeated words: "Scott likes eggs. Scott is allergic to eggs" becomes "Scott likes eggs, but he is allergic to them."

One technique for revising writing to increase sentence fluency is "padding" short, simple sentences by adding phrases that provide more details specifying why, how, when, and/or where something took place. For example, a writer might have these two simple sentences: "I went to the market. I purchased a cake." To revise these, the writer can add the following informative dependent and independent clauses and prepositional phrases, respectively: "Before my mother woke up, I sneaked out of the house and went to the supermarket. As a birthday surprise, I purchased a cake for her." When revising sentences to make them longer, writers must also punctuate them correctly to change them from simple sentences to compound, complex, or compound-complex sentences.

Skills Writers Can Employ to Increase Fluency

One way writers can increase fluency is by varying the beginnings of sentences. Writers do this by starting most of their sentences with different words and phrases rather than monotonously repeating the same ones across multiple sentences. Another way writers can increase fluency is by varying the lengths of sentences. Since run-on sentences are incorrect, writers make sentences longer by also converting them from simple to compound, complex, and compound-complex sentences. The coordination and subordination involved in these also give the text more variation and interest, hence more fluency. Here are a few more ways writers can increase fluency:

- Varying the transitional language and conjunctions used makes sentences more fluent.
- Writing sentences with a variety of rhythms by using prepositional phrases.
- Varying sentence structure adds fluency.

Analytical Writing Questions

Analyze an Issue

There are two analytical writing questions, "Analyze an Issue" and "Analyze an Argument." You will have thirty minutes to complete each task. The Analyze an Issue task asks you to think critically about a general issue and express your thoughts on it. In your essay, you will be asked to offer multiple perspectives on the issue and whether you agree or disagree with it, or what position you choose to take on the issue. You will be presented with one issue statement followed by a set of instructions.

The following is the "Analyze an Issue" prompt. Set the timer for thirty minutes, then carefully read over the prompt two or three times. Once you have a grasp on what the question is asking of you, begin writing. Remember to use the knowledge of writing in our content section. The prompt is as follows:

> Government assistance programs such as Medicaid and Children's Health Insurance Program (CHIP) that provide low-cost medical benefits to low-income adults, children, seniors, pregnant women, and people with disabilities, should be federally discontinued and placed in the responsibility of the states.
>
> Write a response in which you agree or disagree with the argument above. Provide your knowledge of the recommendation, and then give your own recommendation and the reasons for it. Explain the benefits that would come from your own recommendation of the issue and how these might shape your argument.

Analyze an Argument

You will have thirty minutes to complete the Analyze an Argument task. The Analyze an Argument task asks you to evaluate an argument provided by someone else. However, unlike the Analyze an Issue task, you will not agree or disagree with the claim or provide your opinion on whether the argument is correct or not. You are simply testing the validity and structure of the argument presented.

When provided with the argument, there are a few things you should keep in mind regarding the validity of the argument:

- Is there an adequate amount of evidence presented in the argument?
- Is the evidence relevant to the claims?
- Are there claims that are left unstated or that are unclear?
- How is the structure of the argument? Does it smoothly transition from one point to the other, or is it missing effective transitions?
- Is there anything the author assumes without proof?

Remember, do not offer your stance on the argument. You should not offer your opinion on whether the argument is true or not, and you should not express your views on the argument. You are simply analyzing the structure of the argument and seeing what could be improved. The prompt is as follows:

> Substance abuse recovery centers in Florida have doubled in the past two years. Additionally, the centers that already existed are filled to the maximum occupancy. In order to accommodate the patients, some recovery centers are expanding their residencies and adding more staff. The reason for this rise in recovery centers speaks to the state of our nation with the abuse of drugs

and alcohol. The substance abuse problem is worse than ever before, and will only continue to grow if we allow the trafficking and consumption of drugs to go unpunished.

Write a response in which you identify the stated or unstated assumptions in the argument. How does the argument depend on these assumptions? Make sure to provide details of what the argument would look like if those assumptions prove unwarranted.

Greetings!

First, we would like to give a huge "thank you" for choosing us and this study guide for your GRE exam. We hope that it will lead you to success on this exam and for your years to come.

Our team has tried to make your preparations as thorough as possible by covering all of the topics you should be expected to know. In addition, our writers attempted to create practice questions identical to what you will see on the day of your actual test. We have also included many test-taking strategies to help you learn the material, maintain the knowledge, and take the test with confidence.

We strive for excellence in our products, and if you have any comments or concerns over the quality of something in this study guide, please send us an email so that we may improve.

As you continue forward in life, we would like to remain alongside you with other books and study guides in our library. We are continually producing and updating study guides in several different subjects. If you are looking for something in particular, all of our products are available on Amazon. You may also send us an email!

Sincerely,
APEX Publishing
info@apexprep.com

Free Study Tips DVD

In addition to the tips and content in this guide, we have created a FREE DVD with helpful study tips to further assist your exam preparation. **This FREE Study Tips DVD provides you with top-notch tips to conquer your exam and reach your goals.**

Our simple request in exchange for the strategy-packed DVD is that you email us your feedback about our study guide. We would love to hear what you thought about the guide, and we welcome any and all feedback—positive, negative, or neutral. It is our #1 goal to provide you with top quality products and customer service.

To receive your **FREE Study Tips DVD**, email freedvd@apexprep.com. Please put "FREE DVD" in the subject line and put the following in the email:

a. The name of the study guide you purchased.

b. Your rating of the study guide on a scale of 1-5, with 5 being the highest score.

c. Any thoughts or feedback about your study guide.

d. Your first and last name and your mailing address, so we know where to send your free DVD!

Thank you!

Printed in Great Britain
by Amazon